DNA

and Your Body

During a long and distinguished career in biomedical science, Professor COLIN MASTERS has gained an international reputation for his research in biotechnology and its component disciplines. He is the author of several books and hundreds of research papers in these areas, has been awarded three doctorates and life membership of the Australian Society for Biochemistry and Molecular Biology, and has received many invitations to speak at international scientific conferences.

DNA and Your Body

What you need to know about biotechnology

COLIN MASTERS

A UNSW PRESS BOOK

Published by
University of New South Wales Press Ltd
University of New South Wales
Sydney NSW 2052
AUSTRALIA
www.unswpress.com.au

First published 2005

National Library of Australia
Cataloguing-in-Publication entry

Masters, C. J. (Colin J.).
DNA and your body: what you need to know about biotechnology.
Includes index.
ISBN 0 86840 984 7.
1. Biotechnology - Popular works. 2. DNA - Popular works.
I. Title.

660.6

Design Di Quick
Cover image Getty Images
Printer Griffin Press

CONTENTS

ACKNOWLEDGMENTS

This book could not have been published without the contributions and assistance of many people.

I would like to thank colleagues with whom I've discussed aspects of this book, staff at UNSW Press for their assistance and support, Stephen Francis for the illustrations, and my wife, Josephine, for her understanding and forbearance during the writing of this book.

To all of the above, I wish to extend my sincere gratitude.

INTRODUCTION

DNA has been described as our ticket to tomorrow – and this book aims to provide an easily understood, ready-reference road map to the major destinations en route.

If you asked a cross-section of the world's adult population, 'what has been the most amazing, most emotional experience of your life?', it's likely that a great many people would answer along the lines of 'being involved with the conception, birth and development of my children'. Both mothers and fathers in all cultures keenly follow the stages of their children's development, prenatal and postnatal, and are engrossed by the wonder of the development process.

Even if we are not parents, most of us have an intense interest in the way the bodies of human beings develop, and a desire to understand the workings of this extraordinary biological process. At the same time, there is a general fascination with DNA, and a widespread appreciation of its central role in genetics, the biology of development and such matters as inherited disease, abnormal development, and the possibilities of cloning and genetic engineering.

The relationship between DNA and our body's development is a matter of immediate interest to many in this day and age, and it is a core theme in this book.

Most people know that DNA is the molecule that programs our human potential; and we are constantly being told that major advances in medicine, agriculture and our general standard of living will flow from current research that is rapidly unlocking its many secrets. We have moved into the Age of Biotechnology, and are confronted on a daily basis with media speculation on the potential significance of the latest DNA discoveries.

Many of these developments may, in fact, be of momentous significance, for our present and for our future. They may even revolutionise the nature of our existence. They certainly raise important questions about the directions in which society is heading – questions that deserve a broad understanding and an informed debate. But are we ready for such a debate in this gene age?

There are dramatic developments in prospect. The flood of new knowledge and new technology is transforming the way medicine is practiced, modifying the foods we eat, and revolutionising our understanding of the ageing process. It is changing the way crimes are solved, the way family relationships are established, even the traditional methods of procreation. The issues have far-reaching implications for our everyday lives; and they won't go away. They are destined to be increasingly prominent in public consideration and debate.

A general understanding of the basic features of DNA science and how it impinges on so many aspects of our life is critical for our future. It has become essential for non-scientists to understand something about the implications of our growing knowledge of DNA, and its relationship to human development, health and well-being.

Only an informed public – not just scientists and politicians – can ensure that DNA technology is used wisely and to the greatest benefit of humanity.

This book seeks to describe the basic principles of DNA, and the central elements of biotechnology, in terms comprehensible to non-scientists. The details may be complex, and the pace of development breath-taking; but the principles are simple enough, and are explained here in the hope that understanding them will enable you to make an informed contribution to the debate of the century.

If you want to know more about any of the topics dealt with you should be able to find what you are looking for in one or more of the publications listed in the *Further reading* section at the end of this book. There is also a glossary of the technical terms used, to help you find your way around; and a list of frequently asked DNA questions is included to give you quick answers to vital questions.

1869	Friedrich Miescher first identifies nucleic acids.
1944	Oswald Avery and his colleagues propose that DNA is the genetic material in most organisms.
1940s	Chemical analyses by Erwin Chargaff and his colleagues show that DNA from all life forms has the same constituents.
1953	James Watson and Francis Crick described the structure of DNA as a double helix, and propose its method of replication.
1962	Watson, Crick and Maurice Wilkins receive the Nobel Prize for their pioneering work on DNA.
1970	The first gene is synthesised in the laboratory.
1986	The first genetically engineered vaccine for humans (for hepatitis B) is developed.
1989	The National Centre for Human Genome Research is formed in the US to oversee the sequencing of human DNA – the human genome project.

1990	DNA, and cloning, enter popular culture with the publication of *Jurassic Park*.
1997	Dolly the lamb, the first cloned mammal, is born.
2000	The first synthetic DNA is produced, followed by the prediction that artificial life forms will be created within a few years. The first designer baby is born.
2001	Results from the human genome project are released.
2003	Dolly is put down, suffering from degenerative disease and 'old before her time'.
2004	Human embryos are cloned and developed to a stage where embryonic stem cells can be extracted.
2005	Public debate intensifies about the ethical issues associated with DNA and its manipulation.

FREQUENTLY ASKED QUESTIONS

I've heard that chimpanzees and humans have around 98% of their DNA in common. Does this mean that chimps are almost human?

Not really! Humans have their own special characteristics, and most of the DNA differences occur in what's called the non-coding DNA. We don't yet know a great deal about this type of DNA, but what we're finding out is pretty exciting. **Discover more on pages 46 and 163.**

The human genome project sounds very interesting, but is it going to serve any practical purpose?

Very much so. The knowledge gained from the human genome project is already revolutionising our understanding of disease, and it promises huge advances in medicine in the future. Chapter 3 explains what the human genome project is all about.

FREQUENTLY ASKED QUESTIONS

Q Now that scientists have cloned the sheep Dolly, couldn't they just as easily clone a human being ?

A **Not quite. Although it is likely that the technology to clone human beings will soon be widely available, at the moment there a number of reasons why cloning a person is much more difficult than cloning a sheep. These obstacles are explained on pages 61 to 62.**

Q I know that there is a lot of controversy about genetically modified foods – some people seem to think they're going to save the world, other people seem to think they're going to destroy it. What exactly are genetically modified foods, and what are the facts?

A **Genetically modified foods come from crops that have had genes inserted into their DNA, sometimes from other species, to improve their performance. The controversy about them is summarised on pages 73 to 76.**

FREQUENTLY ASKED QUESTIONS

Is it possible that in the future we'll be able to grow ourselves new organs to replace those damaged by disease, or is this science fiction?

It's probably only a matter of time. The state of research on growing new body parts is discussed in chapter 6.

Scientists seem to spend a lot of time looking at insects and bacteria. How is investigating the DNA of a fruit fly going to help us understand human disease?

Because all life forms have the same basic constituents in their DNA, it's possible to get information about a complex organism by studying parallel structures and functions in a simple one. **Find out more on page 151.**

I've heard that some companies have taken out patents on genes. How can they do that? How can anyone own our DNA?

There is a lot of controversy surrounding the ownership of genetic material, and you may find some of the legal decisions surprising. **Read about them on pages 49 to 50.**

Q Is there a fountain of youth? Is it possible for humans live to be 150 (or 1500) if we just keep on increasing our knowledge of how the body works, and get better at fixing it when it goes wrong?

A **The fountain of youth hasn't been discovered yet, and scientists have differing views about whether it ever will be. In the meantime, there are definitely things we can do to improve our health and longevity. The relationship between DNA and ageing is discussed in chapter 8.**

Q Why would anyone need to understand DNA and its role in biotechnology? Aren't the scientists taking care of it?

A **Recent advances in DNA technology have tremendous significance for our society and our future, and only an informed public is going to be in a position to make appropriate decisions about them. We simply can't afford to leave it to the scientists alone.**

CHAPTER 1

DNA, GROWTH AND DIFFERENTIATION

The DNA molecule contains the chemical guidelines that steer the development of the human body from embryo to adulthood. Understanding how it does this – as well as its roles in medicine and biotechnology – requires a basic understanding of the main elements of its structure.

DNA in a fertilized egg replicates itself over and over, and is eventually present in identical form in millions of cells in the adult human body. The DNA carries the information necessary for the fertilized egg to differentiate into the body's many hundreds of cell types, and for protein synthesis – the process by which those cells make the proteins necessary for life.

This chapter introduces the structure of DNA, and describes the fundamental mechanisms of growth and cellular differentiation.

It is now just over half a century since James Watson and Francis Crick published their description of the structure of DNA. Since then the ability of this molecule to self-replicate – that is, to use surrounding materials to make exact copies of itself – has become a cornerstone of modern biology and medicine. Some writers have even gone so far as to proclaim Crick and Watson's solution to the puzzle of DNA structure the most significant discovery in human history.

We now know that DNA is the genetic material of all cellular organisms (animals, plants and bacteria) and many viruses; and that besides having the capacity to replicate itself, it is the source of information for the synthesis of some of the most important functional molecules – proteins – in all living organisms. The DNA molecule also contains the chemical guidelines that steer the development of our bodies from the moment of conception, through the embryonic stages and infancy, all the way to adulthood, and in so doing may give a forecast for our future.

To understand this extraordinary capacity – and to understand much of what follows – we need to take a brief look at the chemical nature of the DNA molecule.

The structure of DNA

DNA stands for *deoxyribonucleic acid*. It is a large, complex molecule that occurs in the cell nucleus of all multi-cellular organisms. (Bacteria and viruses don't have cell nuclei, but we don't need to concern ourselves with them at this stage).

Besides the DNA in the nucleus, human cells also contain a small amount of DNA outside the nucleus, in small *organelles* called *mitochondria*, which are mainly concerned with providing energy for the cell. We'll have more to say about the mitochondria in chapter 2. Figure 1.1 shows the

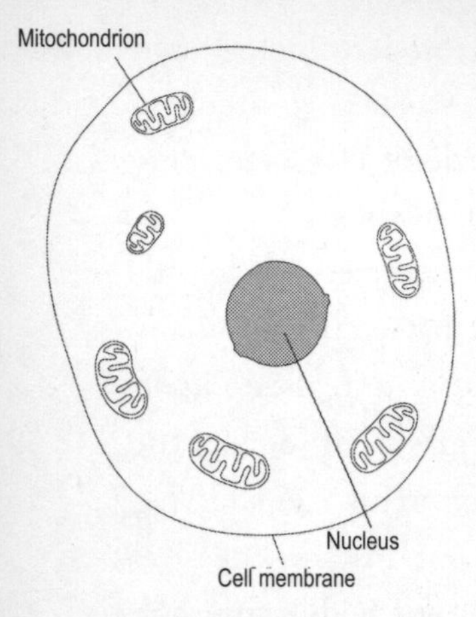

FIGURE 1.1 Cell showing the location of the nucleus.

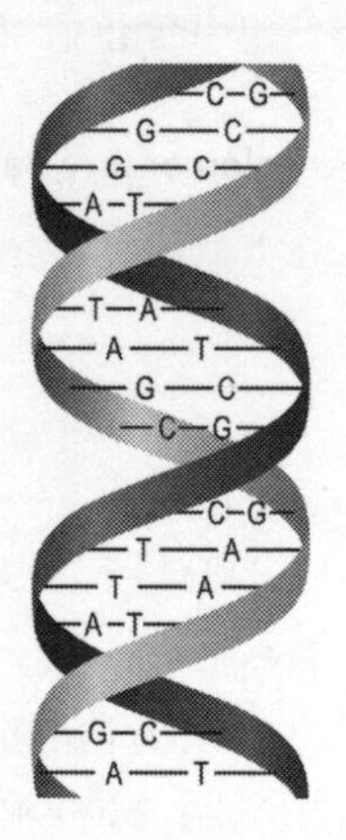

FIGURE 1.2 The double-helical model of DNA. The two ribbons represent the sugar and phosphate part of the molecule. The steps labelled A–T and G–C represent the adenosine–thymine and guanine–cytosine base pairs.

basic structure of a cell, with the nucleus and the mitochondria.

The form of the DNA molecule is often described as a *double helix.* You may already be familiar with the general structure shown in figure 1.2.

Put simply, the DNA molecule consists of two strands of chemicals called *nucleotides*, each of which consists of a *base*, a sugar and a phosphate. The molecule looks a bit like a spiral staircase; the 'banisters' of the staircase are made from the sugar and phosphate molecules, and the 'steps' are pairs of bases.

The two nucleotide strands correspond to each other in a quite specific way: the base adenine (A) is *always* paired with the base thymine (T) to form one type of step, while the base cytosine (C) is *always* paired with the base guanine (G) to form another type.

DNA is a generic or collective term. The general structure just described allows enormous scope for individual variation through changes in the sequences of the base pairs. This means that, although the genetic information of all individual human beings is carried by DNA, each individual person's DNA is different from anyone else's.

In animals, DNA is organised into DNA–protein complexes called *chromosomes*, and sections of DNA with particular functions are called *genes*, with many genes being carried on each chromosome. We will have a lot to say about chromosomes and genes in later chapters.

A nucleotide is an organic molecule consisting of a base, a sugar and a phosphate.

Growth and the transfer of genetic information

The structure of DNA put forward by Watson and Crick gave a prophetic insight into how genetic material is replicated in the cells of humans (indeed, of all animals and plants). The fact that the nucleotide pairings must always be the same enables a simple yet elegant mechanism for replication. As the two strands of the DNA molecule separate in a prelude to cell division, they provide a template for the formation of an exact copy of the original molecule, and of the genetic information that it carries. The chemicals needed for the new strands are obtained from stores within the cell. This mechanism is illustrated in figure 1.3.

A base pair consists of two nucleotides that are linked through their complementary bases.

Humans begin their existence as a single cell, a fertilized egg, which contains DNA from both a mother and a father. During its subsequent growth this cell divides many times – enough times to account for the presence of a hundred million million cells in the adult human body. With only a few specialised exceptions, all the cells in any particular human body contain exactly the same sequence of

DNA is composed of long strands of nucleotides, generally in pairs.

nucleotides in their DNA. (The exceptions are the *germ line cells* – cells containing genetic information that may be passed on to offspring, including egg and sperm cells.)

The reason that DNA differs from one person to another is that the original fertilized egg contains DNA contributed by two parents. Because of the very considerable scope for variation in the resulting mixture, DNA differs to some degree between siblings (except in the case of identical twins), and to a far greater extent between individuals who are not closely related.

This characteristic property of DNA – that its composition is the same in the cells of any one person, but differs between individuals – has a profound significance for development and differentiation, and other aspects of biology.

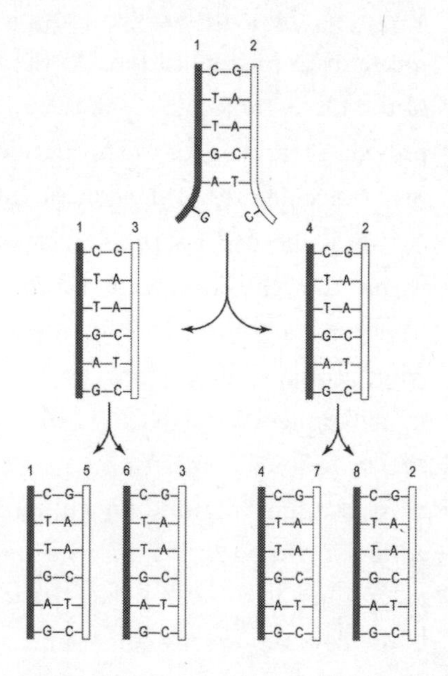

FIGURE 1.3 DNA replication. For simplicity, the strands are shown as lines. As the parent strands (1, 2) separate, they expose their complementary base sequences. Since each base will only combine with one other particular base, the base sequences of the new molecules (1–3 and 4–2) are exactly the same as those in the original DNA. This identity of structure is maintained through successive generations (1,5; 3,6; 4,7; 2,8; etc.)

DNA and detectives

For many decades, the principal means of determining that a person had been present at a crime scene was fingerprinting. It is well known that this technique works on the basis that the fingerprints of an individual are unique (or very nearly so), so finding a person's fingerprints in a given location provides solid evidence that that person was there. All readers of detective stories, or followers of TV cop shows, will be familiar with this approach.

The developing technique of DNA analysis has proved to have many advantages over fingerprinting, and is now largely superseding the older technique in police work. Indeed, DNA profiling is generally recognised as the most important advance in crime detection in the last hundred years. Like fingerprinting, DNA analysis is based on the uniqueness of something a person leaves behind; but in this case, the something is the person's genetic material.

As we have seen, the individual cells of a particular person have the same DNA – the same sequence of nucleotides, derived by successive replications from the DNA of the original fertilized egg during cell division and multiplication. So any cellular material left behind at the scene of a crime may provide firm evidence of identity. The smallest trace – even a single cell from blood or skin, say – may be used for this purpose through a laboratory process that amplifies the amount of DNA until there is enough for the clear identification of the nucleotide sequences it contains. >

> DNA 'fingerprinting' has provided forensic science with a powerful investigative technique that requires minimal material, is extremely specific, and is more widely applicable and less susceptible to cover-up or removal than fingerprints.

DNA labelling

Other uses have been made of the individuality of the human genetic code. At the 2000 Olympic Games in Sydney, in a high-tech effort to combat the sale of counterfeit merchandise, official souvenirs were tagged with the DNA of an Australian gold medallist so that they could be readily identified as genuine.

A similar technology has been developed to authenticate fine art – a short strand of a person's DNA is mixed with invisible ink, and the added security of a luminescent chemical that is only visible with the aid of a special decoder. This provides artists with a foolproof method of authenticating their works. DNA was also used to label the baseball with which Mark Maguire hit his record-setting seventieth home run, a ball that fetched $US 3 million at auction.

It has even been suggested that a similar technique should be used to make paper currency more difficult to counterfeit.

DNA and taxonomy

Differences in DNA composition also have practical applications in *taxonomy* – the classification of living things. The technique has been widely used to determine evolutionary relationships between animals, plants and other life forms, for example. The guiding principle is that the more closely related species are, the more similar will be their DNA.

DNA and protein synthesis

As well as providing a mechanism for its self-replication, DNA carries the information necessary for *protein synthesis*, the mechanism that produces most of the functional material – proteins – in our cells and tissues, and is mainly responsible for the multitude of different cell types in our bodies.

Essentially, DNA in the nucleus of a cell transcribes its genetic information to another form of nucleic acid called *messenger RNA* (mRNA) by a process similar to that of DNA replication. Because mRNA is a much smaller molecule than DNA, it can move out of the nucleus through pores in the nuclear membrane to other parts of the cell and initiate protein synthesis where it is required. Figure 1.4 gives a simple illustration of the initial step in this process.

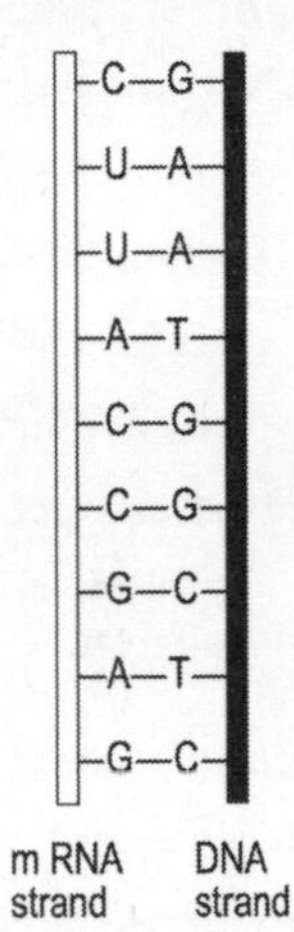

FIGURE 1.4 The initial step in protein synthesis. The basic information for protein synthesis is transferred from one of the DNA strands to mRNA by the formation of complementary base sequences. RNA has a different base composition from DNA, with uracil (U) replacing thymine (T).

DNA and cellular differentiation

Early development

As we have seen, the early development of human beings (and many other organisms) begins with a fertilized egg. When a sperm enters an egg it initiates a wave of *calcium ions*, which starts the process of development. The fertilized egg undergoes *cleavage* – first into two cells, which cleave into four, then into eight, sixteen, and so on – all the time replicating the original DNA, so that the DNA in all the cells of the developing animal is exactly the same.

When the embryo consists of a hundred or so cells, there is a change. Whereas all the cells were initially identical in appearance and composition, they now begin to assume individual characteristics in a process called cellular *differentiation.*

In most species, including humans, the solid mass of cells now forms a hollow sphere, the *blastocyst.* This becomes a double-walled sac, the *gastrula*, with an outer wall called the *ectoderm* and an inner wall called the *endoderm.* A third layer, the *mesoderm*, develops between them. These three layers are known as the *primary germ layers*, and they develop into the same type of organs in all animal species. The endoderm produces the lining

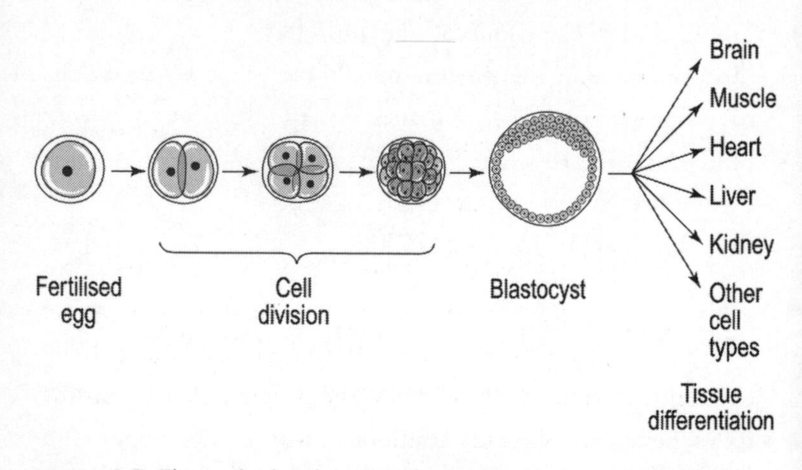

FIGURE 1.5 The early developmental sequence in humans. Following fertilization, the egg divides and the cells multiply until they reach the blastocyst stage, which implants in the uterus. Differentiation begins and eventually produces all the cell types.

of the air passages and most of the digestive system; the mesoderm gives rise to blood, blood vessels, muscle, connective tissue and many of the internal organs; while the ectoderm develops into the epidermis (skin), central nervous system and mucous membranes.

There's no need to remember all these technical terms at the moment – although some of them may be familiar to you from their increasingly frequent use in press accounts of DNA biology. Just visualise the sequence of events illustrated in figure 1.5.

Cellular differentiation

During differentiation a diversity of cell and tissue types develops: first in the embryo, then increasing in complexity in the foetus and young animal through to the adult. By any standards, it's an impressive transformation. In the adult human, for example, there are some 100 000 000 000 000 (a hundred million million) cells and hundreds of different cell types.

> Differentiation is the process whereby the cells of the early embryo develop into the different cell types of the adult body.

What is the role of DNA in these spectacular cellular transformations? We know that DNA provides the template or genetic information for developmental changes; but, as we have seen, almost every cell in the individual human body contains the same DNA. How do the differences between the different cell types come about?

First, we need to consider what these differences are.

The main compositional difference between cell types may be traced to the number and type of proteins they contain. For example, the principal protein in red blood cells is *haemoglobin*; and the presence of haemoglobin in substantial quantities in a cell is sufficient to characterise that particular cell type. Muscle cells are characterised by their high content of *contractile* proteins; tendons and cartilage contain mainly *structural* proteins; the beta cells of the pancreas contain *insulin*, and so on.

We can generalise by saying that:

- each cell type has its own characteristic make-up of proteins
- these proteins are functional elements in the cells (that is, they are necessary for all the essential cell functions)
- protein composition is the most common analytical tool used to distinguish between cell types.

Proteins and genes

The section of DNA responsible for the synthesis of a particular protein is called a *gene*, and a human being has around thirty thousand of them. DNA establishes the different and characteristic patterns of proteins in the different cell types through a process called *differential gene function*, which involves the selective activation of particular genes in the DNA. This concept is central to an understanding of the process of development and differentiation.

A gene is a section of DNA that contains the code for a particular protein.

The mechanisms of protein synthesis are complex, and since the purpose of this book is to present the basics in a readily understood way rather than to describe the process in scientific detail, we will satisfy ourselves for the present with the statement, once part of the central dogma of molecular biology, that *one gene makes one protein.* We will have more to say about this in chapter 10.

In an adult human body there are tens of thousands of different proteins, so the possibilities for different combinations of proteins in a cell are enormous – more than enough to allow for the different patterns of proteins contained in the hundreds of recognised cell types.

The differences in cell types and tissues in the human body are largely due to the sequential activation of particular genes in the human genome (the total genetic information for a human being) during growth and development.

This process of development and differentiation can be compared to the playing of a long and complex composition on a piano with thirty thousand keys – the genes – where each note represents a molecule of a particular protein. There is immense scope for variation between the different themes – the cells – in the overall composition – the human body.

CHAPTER 2

DNA AND DISEASE

Several thousand known medical conditions are included in the category of genetic disease, ranging in severity from diseases that are fatal in infancy to those that are relatively benign and long lasting. Millions of people spend a lifetime suffering from one or another of these conditions.

These diseases may be classified into several categories – single gene disorders, multifactorial disorders, chromosomal disorders, and diseases affecting the mitochondrial DNA.

Because these genetic diseases are so diverse and widespread, their diagnoses and treatment have become a major concern of medical biology. Vastly improved techniques have allowed faster and more specific diagnosis in many cases. Screening can now be carried out during most stages of pregnancy and through to adulthood. This chapter gives examples of a number of specific genetic diseases, and discusses some of the ethical issues involved in the diagnosis and treatment of genetic disease.

We have seen how DNA directs development and tissue differentiation in humans and other animals. If all goes well, this results in a normal healthy adult. If there is any alteration or damage to the DNA, however, normal development may be affected, and disease may develop. In fact, abnormalities of this type are a widespread and significant feature of human life.

Some ten thousand genetic disorders have been identified, ranging in severity from diseases with fatal consequences to those with relatively innocuous effects, such as colour blindness. They all involve alteration to one or more of the genes in a person's DNA.

We will consider first disorders involving only a single gene, and then move on to those that are more complex.

Single gene disorders

As the name indicates, these disorders result from errors in a single gene. A few of these are shown in the box on page 24.

Mutations

Changes in the structure of genes are called *mutations*. Mutations can occur, for example, during cell replication, the process by which a cell splits itself into two identical copies. Each new cell should receive an exact copy of the DNA from the original cell – occasionally, however, errors occur, and a gene is changed. This may affect the gene's particular protein product, which may be produced in insufficient quantities, or in an altered form. In either case the protein may be unable to carry out its function, and a genetic disorder may be the result.

A mutation is a permanent change in the structure of a gene. Many are harmless; others cause diseases of varying degrees of severity.

SOME SINGLE GENE DISORDERS

DISEASE	EFFECTS
cystic fibrosis	chronic pulmonary disease, generally leading to death at an early age
Lesch-Nyhan syndrome	self-mutilation, mental retardation
Gaucher's disease	erosion of bones and hip joints; sometimes brain damage
gout	recurring attacks of acute arthritis
hypercholesterolaemia	the familial form often leads to heart disease, and sometimes to early death from heart failure
rickets	short stature, convulsions
sickle-cell anaemia	can lead to severe pain in joints, and death
Tay-Sachs disease	mental deterioration, death at an early age

Some mutations occur spontaneously. Others are caused by factors in the environment known as *mutagens*. Examples of mutagens that affect human DNA include ultra-violet light, X-rays and various chemicals. While mutations are common (and often harmless), an error in the order of even one base pair can cause serious medical problems if it disrupts a critical gene. Single gene disorders occur in some five per cent of live human births, for example, and contribute to some twenty per cent of deaths occurring in the first one or two years of life.

A frequently cited example of a single gene disorder is *sickle-cell anaemia*. In this condition, a mutated gene produces an altered form of haemoglobin – the main protein of red blood cells, whose primary function is to transport oxygen from the lungs to the other tissues in the body. This is a vital function, since metabolism in our tissues depends on an adequate and continuous supply of oxygen. In sickle-cell anaemia, this function is partially and painfully deficient.

Recessive gene disorders

Once a mutation has occurred in a particular gene, the mutated gene is passed on to descendents in exactly the same way as a normal gene.

Genes occur in pairs (except for *X-linked* genes – more about these later), with one of the pair being inherited from the mother and one from the father. In many cases disease only develops if a person inherits faulty copies of a particular gene from both parents. The gene in this case is called *recessive*. In some single gene disorders a person carrying only one copy of the faulty gene has no symptoms at all. In others, the person has much milder symptoms than if they had two copies. Sickle-cell anaemia is an example. Individuals who have a single faulty gene are said to be *carriers* of the disease. Most of their haemoglobin is in the normal form, and consequently the symptoms are much milder than in full sickle-cell anaemia.

Sickle-cell anaemia affects hundreds of thousands of people throughout the world. A reason for this high frequency of occurrence is that the carrier state, which produces relatively mild clinical symptoms, also gives improved protection against malaria. Hence there is a selection pressure to maintain the faulty gene in populations living in malaria prone regions.

Another example of a recessive single gene disorder is *cystic fibrosis*, one of the most common fatal genetic diseases in many western countries. The condition involves a protein that is normally present in cells that line the inner surface of the lungs, and helps transport chloride molecules across cell membranes. In cystic fibrosis, thick and suffocating mucus builds up in the lungs, resulting in infections that can be lethal. At present, half of those who suffer from cystic fibrosis die before the age of thirty.

To develop cystic fibrosis, a person must inherit the defect from both parents. One in thirty of the population of many countries carries the defective gene. If both members of a couple carry the gene, each child they have has a twenty-five per cent chance of having the condition.

Dominant gene disorders

In other cases of single gene disorders, the faulty gene exerts its full effect when only one copy is present. The gene in this case is called *dominant*. An example of this type of condition is *Huntington's disease*. In Huntington's disease, the person usually does not exhibit symptoms until mid-life (at the age of around thirty to forty). At this stage, the person begins to develop problems with coordination, thinking and judgement – symptoms of a degeneration of nerve cells, which typically results in premature death.

MUTATIONS CAUSING PARTIAL LOSS OF FUNCTION

In sickle-cell anaemia, the mutation causes only partial loss of the normal function of haemoglobin. Total loss of an essential function such as this would result in death. In other cases, the protein's function is not essential to continued life, and total loss of function results in a disorder rather than immediate death, although premature death may be the ultimate outcome. Many of the single gene disorders fall into this category. *Tay-Sachs disease* is one example. People with this disease lack an enzyme (hexosaminidase A) that normally breaks down certain fatty substances in the brain and in nerve cells. Without the enzyme these substances gradually build up in the central nervous system, eventually causing loss of function and, finally, death. Typically this condition becomes evident in the first half year of life, with death occurring before four years of age.

The faulty gene must be inherited from both parents for Tay-Sachs disease to occur.

X-linked disorders

In another category of single gene disorder – the X-linked disorders – the faulty genes are located on what is called the X chromosome. In humans and most other mammals, females carry two copies of the X chromosome, while males carry one copy of the X chromosome paired with a single Y chromosome. Males are at greater risk of these disorders, because they do not carry a second X gene that might be able to compensate for the imperfection.

An example of an X-linked disorder is *haemophilia*. People with this condition lack a normal blood clotting factor, and experience excessive bleeding from minor cuts and bruises. Most people with haemophilia are males – only females who have inherited faulty genes from the X chromosomes of both parents suffer from this condition.

Chromosomal disorders

The genes of plants and animals are arranged on rod-like structures called *chromosomes*. There are normally 23 pairs of chromosomes in the nucleus of each human cell.

The defects in DNA that cause genetic diseases may often be traced back to particular chromosomes. There is now a long list of such associations. Some of these disease–chromosome relationships are shown in the box to the right.

In addition to single defects of the DNA, some chromosomal disorders are characterised by an abnormality in the number or the structure of the chromosomes. As might be expected, these disorders can produce multiple clinical symptoms, which often include a degree of mental retardation.

Chromosomes are rod-like structures in the cell nucleus consisting of a DNA molecule and a number of proteins. Genes are carried on the DNA part of the chromosomes.

The most common example is *Down's syndrome*, in which the affected person has three copies of a particular chromosome (number 21) instead of two. People with Down's syndrome always have some degree of learning difficulty, and may also have heart defects and other serious health problems.

THE CHROMOSOMAL POSITIONING OF SOME GENETIC DISORDERS

chromosome 1	prostate cancer, deafness
chromosome 2	hypothyroidism, colorectal cancer
chromosome 3	dementia, HIV susceptibility
chromosome 4	Huntington's disease, polycystic kidney disease
chromosome 5	endometrial cancer, spinal muscular atrophy
chromosome 6	haemochromatosis, schizophrenia
chromosome 7	cystic fibrosis, dwarfism
chromosome 8	haemolytic anaemia, Burkitt's lymphoma
chromosome 9	cardiomyopathy, fructose intolerance
chromosome 10	Cockayne syndrome, congenital cataracts
chromosome 11	sickle-cell anaemia, albinism
chromosome 12	rickets, bowel inflammation
chromosome 13	breast cancer, pancreatic cancer
chromosome 14	leukaemia, goitre
chromosome 15	Marfan's syndrome, juvenile epilepsy
chromosome 16	polycystic kidney disease, gastric cancer
chromosome 17	muscular dystrophy, Alzheimer's disease
chromosome 18	carpal tunnel syndrome, diabetes mellitus
chromosome 19	myotonic dystrophy, malignant hyperthermia
chromosome 20	Creutzfeldt-Jakob disease, growth hormone deficiency
chromosome 21	amylotrophic lateral sclerosis, polyglandular disease
chromosome 22	giant cell fibroblastoma, Ewing's sarcoma
X chromosome	colour blindness, haemophilia
Y chromosome	gonadal dysgenesis

Other types of genetic disorders

Multifactorial disorders

Multifactorial disorders may involve the combined influence of several genes along with environmental factors such as diet. One example is *spina bifida*. A person with this condition is born with an opening in the spine that requires surgery, and may have problems developing the ability to walk or to exercise bowel control. While genes play a role in the initiation of damage to the spinal cord, the risk of this defect is significantly increased if the mother does not have adequate amounts of the vitamin folic acid in her diet during early pregnancy.

The range of multifactorial disorders is extremely wide. Many common adult onset disorders such as coronary heart disease, stroke and schizophrenia fall into this category, and genetic analysis of these conditions is one of the most active and important fields of medical genetic research at present.

Mitochondrial disorders

Besides the DNA in the nucleus, human cells also contain a small amount of DNA outside the nucleus, in small structures called mitochondria. These are mainly concerned with the provision of energy for the cell.

Mitochondrial DNA is different from that in the chromosomes, and is passed on in a different way. Both sperm and eggs contain mitochondria at fertilization, but sperm contribute only their nuclear DNA to the fertilized egg. As a result, all of a person's mitochondria come from their mother

– they are all derived from those that were present in the egg before fertilization. So mitochondrial disorders are inherited solely from a person's mother. Both males and females can be affected, but a man who has a mitochondrial disorder will not pass it on to his children.

An example of this type of condition is *Leber's optic neuropathy*, a disorder in which the optic nerve degenerates and causes problems with vision.

Screening and diagnosis

As we have seen, thousands of genetic diseases have been identified so far. These diseases are widespread throughout the world, and although most of them are comparatively rare (occurring in less than 0.1 per cent of the population), they usually have a lifelong effect on those who suffer from them. Genetic disease is thus a major medical issue, and it has presented scientists with considerable challenges in diagnosis and treatment.

Generally, the more health professionals know about an individual's DNA, the better placed they will be to deal with any problems. Clinical geneticists and other health professionals are increasingly using screening tests for genetic disorders, and a number of these are discussed below.

Prenatal screening

Screening during pregnancy may be used to detect foetuses at risk of genetic disorders. This may involve:

- testing the mother's blood for substances that might indicate problems in the foetus

- *chorionic villus* sampling, which involves extracting and examining a sample of the *chorionic villus* (part of the placenta) at about the tenth week of pregnancy
- *amniocentesis*, which involves sampling the amniotic fluid (the fluid that surrounds the foetus in the uterus) between 15 and 20 weeks into the pregnancy. Foetal cells in the fluid are analysed for chromosomal normality, a range of DNA mutations and other indications of genetic disorders.

Both chorionic villus sampling and amniocentesis are invasive, and unfortunately result in miscarriage in about one per cent of cases. New techniques are being developed in which the detection of potential genetic disorders is as simple as taking a maternal blood sample. After the eighth week of pregnancy, cells from the foetus may be separated out from the mother's blood and tested for genetic and chromosomal abnormalities. Because the process is simple and safe, this type of prenatal screening could be offered to all pregnant women, not just those considered at risk as at present.

Preimplantation diagnosis

Preimplantation diagnosis is done during the process of *in vitro fertilization* (IVF). In this procedure the egg is combined with sperm, and goes through the early stages of embryonic development, outside the mother's body. It is possible to remove a cell from the developing embryo before it is implanted into the mother, and test its DNA for specific genetic disorders.

Newborn screening

Probably the most widespread genetic screening takes place immediately after birth, through blood or urine tests. All babies are routinely investigated in many countries. The process is used to detect treatable conditions with potentially serious consequences such as *phenylketonuria* and *galactosaemia*.

Adult screening

Genetic screening may be used to identify carriers of genetic disease – people who carry a single copy of a mutated gene but show no symptoms themselves. Such tests may help to determine whether a couple is at risk of having a child with the disease.

The advantages of DNA diagnosis

Using DNA for diagnosis has major practical advantages over conventional methods. In particular:

- any cell in the body that has a nucleus is suitable for testing because DNA is identical in all these cells
- because a person's DNA is the same whatever their age, it is possible to make a genetic diagnosis before symptoms develop. Predictive testing can identify those at risk, and reassure those who do not carry a mutated gene.

Screening technology – DNA chips

DNA testing comes at a price. It requires an enormous bank of information on the structure of genes in normality and in disease. Fortunately much of this is at hand, or coming to hand, through the *human genome project* (discussed in chapter 3), and there have also been recent improvements in analytical

SCREENING FOR CYSTIC FIBROSIS

One of the more common genetic disorders, cystic fibrosis, can only affect a person who inherits a gene mutation from both parents. When both members of a couple carry the mutation, there is a one in four chance that any particular child of that couple will get the disease.

If either member of a couple has a family history of cystic fibrosis and the woman is pregnant or wishes to become pregnant, both partners may be tested for the mutation. In most populations, about three quarters of the mutations are due to one particular defect in the DNA, but tests may also need to be done for other possible mutations. If both members of the couple are found to carry the gene, they may decide not to have children; or, if there is already a pregnancy, they have the option of prenatal diagnosis, generally through a chorionic villus sample or an amniocentesis. The presence of two gene mutations of the type that causes cystic fibrosis will indicate that the foetus has the disorder, in which case the parents may decide to terminate the pregnancy.

methods. DNA chips are now available, for example, that can detect fatal hereditary disease in embryos produced by IVF techniques. These chips carry on their surface synthetic single stranded DNA sequences that are identical to a normal gene. A sample of embryo DNA can be bound overnight to the synthetic DNA, and any mutations tagged with a special dye and identified by a scanner.

Such techniques could bring about a revolution in the methodology of diagnostic medicine. Doctors could use these chips to examine the genes of even healthy people of any age, establish whether they were at risk of a particular disease and begin early preventative treatment. It is possible to analyse tens of thousands of gene sequences in a single test by this method, and detect tiny differences in genetic sequences. This information could enable doctors to decide on the most effective therapies for any condition to which a person was genetically predisposed, and avoid adverse drug reactions.

Treating genetic disorders

Genetic disease is extremely diverse, and there have, in the past, been difficulties in early diagnosis. DNA science has led to considerable improvements in this regard. It also shows tremendous potential for improvements in treatment. For the moment, many of the standard treatments still involve a special diet, or replacement of a missing substance. Children with phenylketonuria, for example, may be treated with a diet free of phenylalanine, and children with cystic fibrosis may be partially treated with pancreatic enzymes in addition to treatments aimed at preventing or curing the associated lung infections.

Gene therapy

In the future many genetic disorders, including fatal disorders, may be cured or prevented by a process that is still mainly experimental – *gene therapy*. This involves introducing a normal gene sequence into a person's DNA to do the job of the faulty gene. The success of this technique depends on:

- an exact knowledge of gene sequences, and
- a means of inserting DNA into the nuclei of human cells.

It has been used so far with only partial success, but its applications are expanding. The technique was first used in 1990 to treat children with a condition characterised by the lack of a protein called *adenosine deaminase* (ADA). A genetically modified virus was used to carry a normal ADA gene into the patient's cells, and acted there to produce this particular protein and restore normal function.

In many ways, successful gene therapy would be the ultimate extension of molecular biology into clinical genetics and medicine. It aims to treat a wide range of disorders by replacing the patient's mutated DNA. Progress continues to be made in this aspect of molecular biology, and most doctors and many other people will probably be exposed to its therapeutic implications in years to come.

Ethical issues

There are many ethical concerns relating to genetic disease and its diagnosis. Some of these overlap with ethical issues mentioned elsewhere in this book, but particular issues that arise in the context of this chapter include:

- the consent of parties to screening
- the availability of appropriate genetic counselling
- the possibility of a degree of stigma associated with the dissemination of results
- the possible use of genetic information by insurers, employers and legal agencies

- the personal dilemmas that may arise from someone's knowledge of their own genetic defects or those of their children.

Genetic testing is unusual among medical processes in that it affects not only individuals, but also their families and society generally. The issues need input not only from experts, but also from society at large.

Views on genetic testing have varied. At one extreme, a Nobel Prize winning scientist was reported as proposing that we tattoo the foreheads of people who carry one copy of recessive, disease-causing genes so that they would not inadvertently have children with another person who carried the same gene. Needless to say, that course of action is not generally favoured today.

A more general concern relates to possible discrimination by employers or the insurance industry if a genetic defect is identified in someone who shows no symptoms of disease. This is not a straightforward situation, because a positive family history may lead to a loaded insurance policy. If DNA testing in these circumstances reveals a normal gene, the family history is not a problem. If a mutated gene is found, however, the person may have insurance difficulties.

There may also be a fine line between research findings and routine DNA diagnosis. The BRCA1 gene offers an example. A mutation in this gene is associated with an 80 per cent lifetime risk that breast cancer will develop. A second gene, BRCA2, has recently been isolated, and it has been suggested that this gene may be even more relevant to breast cancer than BRCA1, which is also associated with ovarian cancer. Very careful assessment is required in a situation involving such complexity and uncertainty. If a pregnant woman is told that her unborn daughter has a mutation in one of these genes, should she consider an abortion? If she goes through with the pregnancy, what does she tell her daughter, and when? How will the child be affected by the knowledge that at some future time she is likely to be stricken with cancer?

These are just a few examples of the significant ethical issues associated with genetic disease. We are on the brink of a new world of genetic knowledge. We need to inform ourselves about the issues, since we will have to make many of these ethical decisions not only as individuals, but as a society.

CHAPTER 3

DNA AND THE HUMAN GENOME PROJECT

The human genome project, and its objective – to determine the detailed gene structure of the human genome – have received much publicity recently, and the billions of dollars that have been directed to it are testimony to its far-reaching biomedical significance.

Technically, the process involves splitting the DNA molecule into smaller, more easily identifiable bits, so that the full sequence of bases in the original molecule can be reconstructed – a massive task by any standards. Automated techniques and computer analyses have allowed much of the sequencing of the human genome to be completed well ahead of schedule, and many of the thirty thousand or so gene sequences have now been specified.

This information has enormous potential in the diagnosis, and eventually the treatment, of human genetic disease. At the same time it raises important ethical and social issues that are still to be resolved.

As we saw in the last chapter, millions of people have genetic disorders of one type or another. What is more, these conditions are usually there for a lifetime – they are not intermittent or short term like most other diseases – so they represent a massive health problem, and there is a corresponding involvement of medical resources.

Fortunately, there is now the promise of greatly improved techniques for the detection and treatment of genetic disorders. These advances depend on a detailed knowledge of the composition of human DNA.

These are the facts that formed the launch pad for the most ambitious project ever undertaken in biomedical science – the *human genome project.* In this chapter we examine the evolution of the project, the genetic basis of the quest, the type of data obtained, the limitations of the data, and the project's ethical implications.

What is the human genome project?

The human genome project is an international scientific collaboration that aims to analyse and determine the structure of human DNA – that is, the complete sequence of the DNA's nucleotide bases. It has been described in terms that indicate its power and potential, such as 'molecular biology's answer to the space program', and 'the instruction manual to the human body.'

The ultimate aim of the human genome project is to connect specific inherited diseases and traits with particular genes. The project will give us a greatly increased understanding of the structure and action of genes and their role in the biochemical processes that underlie many human diseases.

It was anticipated that striking advances in the prevention and treatment of genetic disorders would result from the project, and that the focus in health care might well shift as a result from the detection and treatment of disease to a process of prediction and prevention. There was a lot of discussion of these issues at several major scientific conferences before the project began. It was generally agreed that DNA sequencing was the most promising line to pursue, and that this angle should be used to promote the project to funding agencies.

The project began in earnest in 1990 in the US. Many other nations joined in, and there are now official human genome research projects that include members from European countries and from Japan. Several private companies are also involved in analysing the DNA of the human genome.

The involvement of so many organisations gives one indication of the magnitude of the project. Another is the fact that the human genome is composed of some three billion base pairs and around thirty thousand genes located on 23 chromosomes. Despite the daunting size and complexity of a project that seeks to determine the exact sequence of all these base pairs, the news is encouraging. Scientists completed the determination of the full sequence of the human genome in 2003.

World leaders were enthusiastic about the significance of the achievement. Former US president Bill Clinton, for example, has been quoted as saying:

> Today we are learning the language in which god created life ... Our children may know cancer only as a constellation of stars, not as a disease that kills and maims ... this is biology's moon shot.

And UK Prime Minister Tony Blair has been quoted as saying:

> Let us be in no doubt about what we are witnessing today – a revolution in medical science whose implications far surpass even the discovery of antibiotics.

Genetic technologies

Identification of the genes associated with genetic diseases such as cystic fibrosis, Huntington's disease and muscular dystrophy represents the first step in developing better genetic diagnostic tests and gene therapies to fight these illnesses. At this point we will have a look at the some of the analytical methodologies used for these purposes, and for other exciting biotechnological achievements as well.

DNA sequencing

DNA sequencing is at the core of the human genome project. DNA molecules are too large and have too many features in common to be specifically identified as whole molecules. So the purified sample of DNA to be analysed is first split (*hydrolysed*) into smaller portions that can be individually identified. The entire sequence of bases is then reconstructed from the pieces. The procedures for doing this are now highly automated, and need not be discussed here.

Hybridisation and the gene chip

Once the sequence of bases is known, a number of other possibilities are opened up. One of these is *hybridisation*. This works on the principle that a single strand of DNA will bind, or *hybridise*, with other single DNA strands that contain complementary bases. (Remember that each nucleotide base will combine with only one of the other bases.) The extent of hybridisation will show whether the DNA fragments are closely complementary, or whether there is a greater or lesser degree of mismatch. Any mismatch can be quantified with a high degree of accuracy.

A number of interesting possibilities follow from this ability to compare different DNAs. It enables, for example, the high speed technology for DNA analysis known as the *gene chip*, which was briefly mentioned in the last chapter. This is a small chip of glass or silicon that contains DNA fragments instead of electronic circuits. Each DNA feature on the chip contains a particular sequence of bases. When the chip is hybridised with a sample, the base sequences of the sample can be determined.

This means that samples of the human genome can be analysed for the presence of a mutated gene, a significant development for the diagnosis of genetic diseases and their potential treatment.

What is more, these gene chips can distinguish between DNA that is active in protein synthesis and DNA that is not. This opens up the possibility of testing the effects of particular drugs on groups of genes – a matter of major importance in the pharmacological treatment of other types of disease as well.

A gene chip (or DNA chip) is a glass or silicon chip that contains fragments of DNA, used to test a sample for the presence of particular genes.

SPECIES SIMILARITIES AND BACTERIAL FACTORIES

Because the four nucleotide bases that are characteristic of human DNA also occur in all other life forms on this planet, analysis of DNA by the hybridisation method can be used to assess the degree of similarity between species, giving insights into their evolution.

But the common occurrence of these nucleotides also has other implications. A particularly significant one is that any type of organism is theoretically capable of receiving a gene from another organism. For example, large amounts of the protein produced by a particular gene in humans can be obtained by inserting the gene into bacterial cells. Bacteria replicate very quickly; in a few hours they may make millions of cells, all containing exact copies of the inserted gene. The protein insulin, needed by diabetics, used to be prepared from tissues derived from cows; it was often of moderate quality and in short supply. Now human insulin of high purity is produced in abundance by bacterial factories, using bacteria that contain the human insulin gene.

Limitations of the human genome project

As exciting as the potential of the human genome project undoubtedly is, there are limitations on the applicability of this type of data. These limitations need to be more widely known, and taken into account in any assessments of the project's overall value.

For one thing, despite all the media hype, the appearance of the analytical data is extraordinarily unexciting. In fact, the composition of DNA must seem to most lay people to be stupendously boring, with its millions of reiterations of the four bases. The short section of a typical sequence shown in figure 3.1 illustrates the point. If the full sequence of the human genome was represented in this way, the data would fill several 1000-page telephone books.

...... AACAGTGTTGGACGTAACAGCACTACA
GGCATCAGGTTTGAACTGACTTGGACGTAACAGG
ACTACAGGTTTGAAGTTAACTGACTTGGA
CGTAACAGGACTACAACGTTTGAACTGGACGTAAT
ACAGGTTAACAGTTGAACTGACTTGGAACGTTACT
GACTAGACTAATACACTAACCTGACTCAGTGA.......

FIGURE 3.1 A small portion of the nucleotide sequence of a gene. An entire gene may contain hundreds or thousands of times the number of nucleotides shown here. Nucleotides are shown by their bases (A = adenine, C = cytosine, G = guanine, T = thymine).

The identity of the component genes is not immediately apparent from the sequence of the bases in this mass of data either – it is only revealed by a substantial degree of further analysis. In fact, although the overall base sequences of human DNA have been determined, only a minority of the genes have been quoted as fully sequenced and associated with specific disease conditions.

The experimental difficulties in translating the information from the human genome project into useful medical data are thus considerable, and a great deal of scientific work remains to be done. Although it is not often emphasised in reports about the project, it remains a fact that a knowledge of the millions of base sequences in human DNA does not immediately translate into information of practical medical utility.

There are several reasons for this.

- Not all the DNA in the human genome represents genes active in protein synthesis. Indeed, a considerable proportion of the DNA in mammals is *non-coding DNA* (it used to be called 'junk' DNA) – highly repetitive sequences that were once assumed to be vestiges of the evolutionary history of a species.
- The data on its own does not show which parts of the sequence correspond to genes, or what exactly the function of any particular gene is. A great deal of sophisticated genetic mapping and biochemical identification is necessary before this can be known. The human genome project, immense though it is, is only a first step in the acquisition of this knowledge.
- There is not just one human genome. As we have seen, one person's DNA is different from everyone else's. The final catalogue of human DNA assembled in the human genome project will be a mosaic from many contributors, and will not correspond exactly to the genome of any one individual. This presents significant problems of interpretation in relation to some diseases.
- There are limits on the usefulness of the data in relation to diseases involving more than one gene, or a combination of

genes and other factors. For example, it is known that many of the major diseases that afflict humanity are caused by the interplay of genes and environmental factors – no matter how protective your genes are against lung cancer, if you smoke heavily for years, the odds are that you will get it. Indeed, in a recent survey of the incidence of cancer in identical twins, environmental or lifestyle effects were found to be more significant than genetic predispositions.

Ethical issues

Breakthroughs in decoding and manipulating the information stored in DNA offer the promise of major benefits – for example, diagnosing and treating diseases, obtaining new medicines, bringing criminals to justice, identifying bodies in mass catastrophes such as that caused by the Indian Ocean tsunami of Boxing Day 2004, and improving food crops and farm animals.

At the same time, genetic technologies challenge us to avoid creating new social or environmental problems. We need to prepare now for ethical and legal controversies.

Privacy issues

There are significant questions about genetic testing, for example. Advances in diagnostic testing can provide extensive genetic information about each person – information that may be crucial in saving lives, or helping people decide whether to have children. At the same time it may pose threats to privacy, and raise new possibilities for discrimination by employers and insurers. There are important questions as to how much of this information should be in the private or public domain.

Changing germ line cells

There is also the possibility of changing human populations through human cloning, genetic changes to embryos, and genetic changes that could be inherited. So far, scientists' increasingly sophisticated ability to manipulate DNA has focused on changing a person's *somatic* cells – body cells other than sperm or egg cells (which are *germ line* cells). Such changes are not passed on to future generations. If gene therapy were used to alter sperm or egg cells the changes would be passed on; although they might help prevent diseases in future generations, any mistakes would have long-term consequences.

Limitations of genetic screening

Medical genetic testing raises challenging issues for both doctors and patients, because many such tests provide statistical possibilities rather than a definite prediction of whether a person will develop the symptoms of a particular genetic disease. A test result may indicate, for example, that a person has a 75 per cent chance risk of developing colon cancer by the age of 65. Appropriate screening tests can then be carried out on the person at risk to identify the disease, if it develops, at its earliest stages, when it is most easily treatable.

Doctors and patients must, however, decide at what age screening should be done, and whether the benefits of early screening are worth the drawbacks of frequent screening. These drawbacks include unnecessary patient anxiety, any adverse side-effect of the screening test itself and, in many cases, cost.

Different problems are posed by genetic screening tests that show a person to be at risk of developing conditions for which there are currently no preventive measures or treatments. Most people would find it devastating to learn that they are at risk of a deadly disease that cannot be prevented by medical measures or lifestyle choices.

Equity issues

Other fears about the technologies revolve around such issues as whether they might be used to alter conditions that are now viewed as part of normal human variability (like height or baldness), or whether the benefits of gene therapy might only be available to people in wealthy countries.

Who owns the genome?

Controversy surrounds the ownership of genetic material. Some people believe that a knowledge of genetic sequences should not be owned or used for profit. Life forms were excluded from US patent law until 1980, but in that year the US Supreme Court ruled that *transgenic organisms* – that is, organisms with modified DNA, like the insulin-producing bacteria described earlier – should be viewed as human interventions rather than as natural life forms.

More recently, research institutions and pharmaceutical companies have sought exclusive rights to research results, and potential profits, from genes sequenced by the human genome project. An example is the genetic mutations associated with the BRCA1 and BRAC2 genes. Patents were awarded to a company on the grounds that it discovered the genes and helped determine the role they play in cancer. Some consumer groups fear that such patents will raise the cost of genetic testing, because the company holding a patent for a particular gene will have a monopoly on products associated with it, including genetic testing products. There are already instances where people must pay thousands of dollars for types of genetic testing that were previously available at no cost.

Even more controversial are patents granted over much longer sections of DNA than are occupied by a single gene. In one case, claims were made over more than 50 per cent of the human genome, and millions of dollars in licensing fees have been negotiated for the use of this genetic information. Even though such patents may involve the so-called junk or non-coding DNA, their implications for routine testing and research into genetic disease are immense, and have caused widespread concern in many countries.

International groups have responded variously to the patenting of genes. Costa Rica, for example, has enacted laws to prohibit the patenting of genes of Costa Rican native species. In contrast, Iceland has become the first country to sell the rights to the genetic code of its entire human population to a biotechnology company. This involves putting detailed genetic, medical, and genealogical information about all Iceland's citizens into databases. The company can use the data in exploring the role of genes in disease, and developing screening tests and medical treatments.

Genetic determinism

One outcome of the advances in genetic knowledge, and the increasing media prominence of these discoveries, is a rising belief in *genetic determinism* – the view that all the characteristics of any organism, including a human being, are determined solely by genes. Certainly the human genome project has uncovered some fascinating facts. It is now clear that the relics of our evolution go back at least 800 million years, and that we are genetically related to all other living organisms – to a surprising degree. Bananas display a 50 per cent genetic similarity with humans, fruit flies 60 per cent, mice 85 per cent and chimpanzees more than 98 per cent.

But there are also major differences. Although the human genome has only about thirty thousand genes, not the hundred thousand that had long been predicted, there are significant differences in the way these genes are organised and expressed in humans. Although humans may have a similar number of genes to lower organisms, they are capable of producing up to ten times as many proteins as these organisms, and they produce them in a more sophisticated way. It now seems that although genes have a dominant role in programming behaviour in simple animals, their influence diminishes as organisms become more complex. In humans it becomes more of a balance between genes and environment – nature and nurture.

Ardent genetic determinism fails to acknowledge that many traits are influenced by environment and by personal experience. Nevertheless, it often surfaces in social policy discussions about such matters as violent behaviour and poor performance at school, with some advocates of genetic determinism devaluing the human characteristics of free will and personal responsibility, and playing down the influence of purely environmental factors such as poverty.

The impact of genetic determinism also extends to the realm of medicine. Women's groups, for example, have expressed a fear that the identification of breast cancer genes is causing doctors to view breast cancer as a disease caused solely by faulty genes, and to ignore the role of cancer-causing agents in the environment. Clearly, it is important to maintain a balanced view in these matters – a view that includes all the determinant influences.

CHAPTER 4

DNA AND CLONING

It has been pointed out that only a very few events, scattered over the centuries, have altered our very notion of what it means to be human. The scientific breakthroughs of Copernicus, Galileo, Darwin, and Watson and Crick come to mind.

Cloning has been described as 'the most significant scientific breakthrough of our time'. It raises important issues, both scientific and ethical. It is not possible to make reasoned judgements on the issues involved in cloning, however, without understanding the precise meaning of the term, the methodologies employed, some of the historical background, and the potential benefits.

Previous chapters have drawn attention to the central role of growth and development in those aspects of biotechnology that most closely affect the human condition. To many people, too, early development is the most extraordinary, and the most interesting, event in the human lifespan. An understanding of this development is also critical to the understanding of genetic disease and to the utilisation of knowledge derived from the human genome project.

In this chapter, another aspect of development is considered – the possibilities raised by technologies that enable the cloning of mammals, and perhaps of humans.

First of all, it is important to understand exactly what is meant by the word. A *clone* is an organism that has the same DNA as another organism. The term is not restricted to human beings or even to mammals – it applies across all forms of life. Nor is it anything extraordinary. Any type of asexual reproduction produces clones. Many single-celled organisms such as bacteria replicate themselves by cloning – the cell divides, so that the new cells have exactly the same DNA as each other, and as the old cell. Plants originating from cuttings taken from a single individual can also be called clones.

A clone is any organism, whatever its origin, that has the same DNA as another organism.

Most of the current interest in cloning has been directed towards the possibility of cloning mammals. Even here, however, the term probably covers more than is generally supposed. Identical twins, for example, originate from the division of a single fertilized egg; so they have exactly the same DNA, and are therefore clones. (Non-identical twins grow from two different eggs, and are not clones.)

Methods of cloning

There are two principal methods of cloning animals: embryo splitting and nuclear transfer. They are shown in figure 4.1. In *embryo splitting*, an embryo with as yet undifferentiated cells

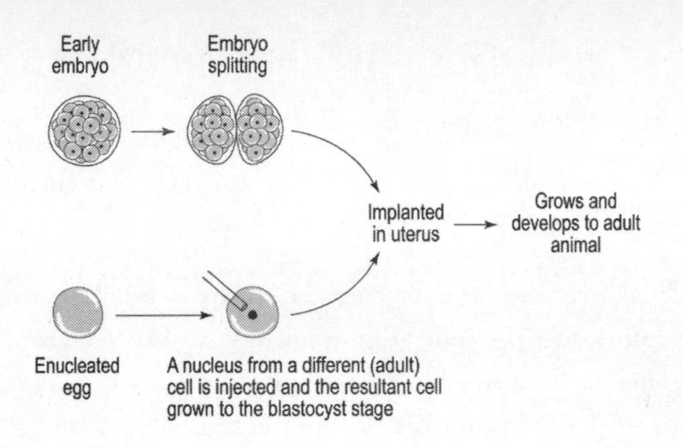

FIGURE 4.1 The main procedures used in the cloning of animals: embryo splitting and nuclear transfer.

is removed from the reproductive tract of the mother and separated into parts, each containing a number of cells. These are implanted into the uterus of another animal, and allowed to develop. The resulting offspring have identical DNA.

Nuclear transfer involves removing an entire nucleus (with its full complement of chromosomal DNA) from a donor cell, and injecting this nucleus into an enucleated egg cell (that is, an egg cell from which the nucleus has been removed). This egg cell may then be capable of developing into an organism identical with the organism from which the donor cell was taken.

Nuclear transfer is theoretically capable of producing large numbers of genetically identical individuals, and it is this technique that has caused such widespread and contentious discussion of the potential of cloning.

To fully appreciate the biological implications of these possibilities, it is useful to consider the history of cloning.

The historical background

In early studies of growth and development, it was observed that differentiation appeared to be irreversible in higher organisms such as mammals.

In the case of plants, it was well known that the whole (adult) plant can be regenerated from a small cutting or sample – clearly indicating that the ability to grow and regenerate is retained in the adult. The situation was quite different with mammals. It was common knowledge that if a limb was severed, it did not grow back – nor could the severed limb regenerate a whole animal. Biologists came to a general conclusion that the differentiation that was characteristic of adult mammals was 'locked in', or irreversible.

Over time, this generalisation came to be questioned. It was observed that a degree of regeneration could take place in some cases. When the liver of an experimental animal had a significant proportion (up to one third) of its mass surgically removed, for example, the liver would grow back to its normal size. Observations such as these raised the intriguing question of whether more complex regeneration might indeed be possible in vertebrate animals. An answer appeared to be supplied by the landmark experiments of John Gurdon at Oxford University.

Gurdon's experiments aroused great interest among biologists, and were recorded in biology textbooks as scientific evidence that the cloning of frogs was possible.

GURDON'S FROGS

Experiments in the US had already demonstrated that cloning was conceptually possible in amphibians. When the nuclei were removed from frogs' eggs and replaced by nuclei from young frog embryos (which had developed to the stage of about ten thousand cells), some of the eggs developed into tadpoles. When the same experiments were tried with nuclei from cells at a later stage of development, however, the success rate plummeted. Whereas about half the cloned cells from the early embryos developed into tadpoles, less than two per cent of cloned cells from tadpoles did so; and the older and more specialized the donor cell, the less likely it was that cloning would succeed.

In short, it appeared that early frog embryos could be cloned, and the DNA of their cells reprogrammed to produce all the tissues of an adult frog; but the more mature the cells from which the DNA was taken (and hence the greater the differentiation that had occurred) the more difficult reprogramming became. It was widely concluded that cloning an adult animal was impossible.

In 1962, Gurdon reported that he had obtained fully developed, mature frogs by transferring the nuclei from the intestinal cells of adult frogs into enucleated frog eggs. It seemed that the cloning of vertebrates was possible after all, and that even the specialized cells of these animals retained the genetic information necessary to direct growth and development in a form that was available to the organism. The DNA in the adult could be reprogrammed to produce all the different types of cells the animal required; differentiation was reversible.

Nevertheless, a degree of scepticism about extending these findings to mammals remained. The experiments in frogs only worked in a very low percentage of cases, the available procedures for transferring the nuclei were relatively crude and capable of causing damage, and some scientists were not convinced that the intestinal cells used by Gurdon were fully differentiated anyway. As late as 1984, a famous scientist was quoted as saying that 'the cloning of mammals by simple nuclear transfer is biologically impossible.'

Dolly

It was not until 1996 that the technical difficulties were overcome sufficiently to allow the successful cloning of a mammal – Dolly, the most famous lamb in history. The Scottish scientists used cells from the udder of a six-year old donor sheep, which they introduced into the enucleated egg of a second sheep. The resultant egg was placed in the uterus of a third sheep, where it developed normally into a lamb.

The key to this achievement lay in understanding the role of the proteins that coat DNA in the nucleus of a cell. These proteins mask the function of as many as 90 per cent of a cell's genes, leaving open only those that the cell needs in order to survive and to perform its specialized functions, as a brain cell or liver cell or whatever. For cloning to succeed the DNA must shed these proteins so that the cell can return to its naked, undifferentiated state, and become the first building block of an entirely new being.

When the scientists introduced the udder cell into the enucleated egg, they jolted it for a few microseconds with a burst of electricity; the pores of the egg and the udder cell membranes opened, and the contents of the udder cell, including its chromosomes, oozed into the egg and took up residence there. The egg acquired the nucleus of the udder cell, and the electric current made it behave as if it were newly fertilized, beginning the sequences of normal differentiation and development. The main features of this process are illustrated in figure 4.2.

Dolly's birth and subsequent growth into an adult represented a major step forward in cloning technology; although strictly speaking Dolly was not a true clone of the sheep from whose cell the donor nucleus was taken. As we know, not all the genes of an animal are found in the cell's nucleus – there

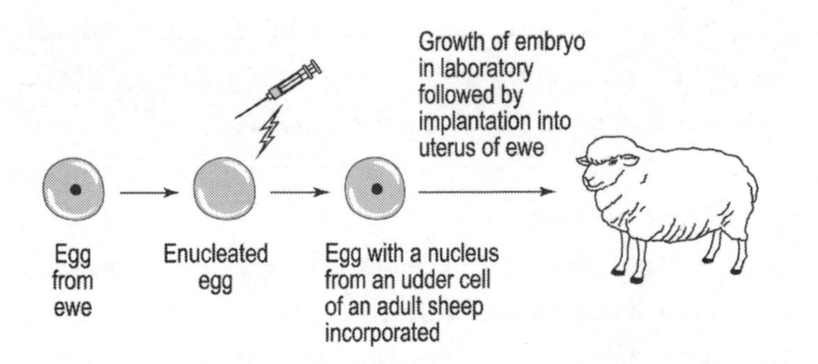

FIGURE 4.2 The major steps in the cloning of Dolly. The nucleus from an adult tissue cell was incorporated into an enucleated egg, then development was allowed to proceed to the blastocyst stage. Following implantation into the uterus, normal development and differentiation occurred, culminating in the birth of Dolly.

are also a few dozen mitochondrial genes located outside the nucleus. The mitochondrial DNA is not transferred with the nucleus, so Dolly in fact contained not only the nuclear DNA from the donor sheep but also the mitochondrial DNA from the sheep that provided the enucleated egg.

The cloning of humans

Dolly was created at the Roslin Institute, an animal research station, and the research was mainly directed towards the possibility of improvements in animal breeding – theoretically introducing a revolution in the field by allowing the rapid production of multiple copies of the best farm animals, and greatly facilitating the generation of many valuable animal products such as *antitrypsin*, a drug obtained from sheep milk. Nevertheless, the news of this momentous achievement inevitably intensified the debate about the possibility of cloning humans. The arrival of Dolly also undoubtedly changed the opinion of many scientists about human cloning from 'impossible' to 'highly probable in the future.'

As to when and how it will occur, the question remains open. There have already been several well-publicised claims of babies being derived from successful human clonings, but as yet none has been scientifically confirmed. And it is clear that many scientists believe that the cloning of human beings is much more complex than the cloning of mammals such as sheep.

Since Dolly, dozens of other animal clones have been produced – cows, pigs, mice, goats and cats. In the case of primates, however, the same procedures have led at the most to the formation of embryos. The cells of primates do not divide normally under the usual cloning procedures, and the embryos have died before developing to the stage of implantation.

One reason for this is becoming apparent. It is known that for the cells of plants and animals to divide properly, their chromosomes must first duplicate themselves, then

CLONING TECHNOLOGY AND FERTILITY TREATMENT

Cloning technology could potentially be used in a fertility treatment to allow women using donated eggs to have a baby that is genetically theirs, rather than the donor's. The technique involves 'gluing' an egg cell from the infertile woman onto an enucleated egg cell donated by another woman, using a protein called phytohaemagglutinin. The two cells are then fused together. (This works because both the cells are egg cells.) The technique would mean that couples using the man's sperm, but another woman's eggs, for in vitro fertilization would be able to have children with a nearly full genetic contribution – with the exception of 37 female mitochondrial genes – from both parents. It would obviously be of help to women whose embryos do not develop because of defects in the cytoplasm (the part of the egg outside the nucleus). This is the case for about ten per cent of the women who have in vitro fertilization treatment.

line up precisely along a structure called a *spindle*. In all mammals, the eggs harbour certain proteins (*spindle forming proteins*) necessary to direct the correct orientation of the chromosomes. In primates, however, these particular proteins are tightly bound to the DNA of the egg. This means that at the first step in cloning, when the host cell is enucleated, the proteins are removed with the DNA, and chromosome reorganisation cannot proceed. In other mammals these special proteins are only loosely bound to the DNA, so that enough remains in the egg cell after enucleation to allow chromosome reorganisation and cell division to take place.

It seems that the cloning of humans will require modifications to the procedures that have been used in the cloning of other mammals, and there are still technical difficulties to overcome.

Under what circumstances might human cloning – the production of live copies of existing individuals – occur? Experiments designed to achieve the duplication of people are illegal in many countries, but not all; and this is due to the fact that the ethical issues of human cloning raise strong passions almost everywhere.

Threats and promises

The threats and promises raised by the achievement of Dolly and other cloning procedures are complex and multilayered. They offer the possibility of real scientific advances that can improve – and save – lives, as well as real dangers and developments that many people would find repugnant.

Potential benefits

In medicine, scientists dream of using cloning techniques to reprogram cells so that we might make our own body parts for transplantation.

GROWING YOUR OWN

Suppose a patient needs a bone marrow transplant. Some deadly forms of leukaemia can be cured completely if the marrow is destroyed and replaced with healthy marrow from someone else. But the marrow must be a close genetic match. If the patient has no close relative whose marrow matches theirs, doctors can search databases of people who have volunteered to donate their marrow, but the odds against finding someone who matches the patient may be as much as a million to one.

An alternative – theoretical at this stage – would be to start a cloning process by taking a cell from the patient and fusing it with an egg cell whose nucleus has been removed. After a few cell divisions, chemicals would be added forcing all the cloned cells (which at this stage of development have the potential to become any part of the body) to become bone marrow cells. The result would be bone marrow cells that were identical with the patient's own marrow.

Anyone who doubts the importance of this need only talk to the families of leukaemia patients, where the only present treatment is likely to be chemotherapy.

In fact, there is a laundry list of potential benefits. The first economically important use of cloning is likely to be the production of replicas of farm animals of high quality, and animals with added genes so that, for example, a cow's milk might contain valuable drugs.

Cloning technology allows scientists to selectively remove genes as well as add them when they create genetically engineered animals. Removal of genes, for example, may be crucial in producing animals whose organs could be used for human transplantation. It is known that one major reason why pig organs are rejected by the human body is the presence of a particular sugar on the surface of the pig organs. If the gene responsible for adding that sugar were removed, the tissues would be more readily accepted.

Potential risks

Of course, the true fascination of the subject still revolves around the questions of why anyone would want to clone a whole human being, and what would happen if we tried.

Many people find this idea morally questionable, and point out that it would be extremely dangerous. No one could reasonably expect that hundreds of human eggs or embryos would be sacrificed for this purpose as sheep embryos needed to be in the Dolly experiment.

It has also been suggested that a clone might age rapidly. Instead of having an expected lifespan of 70 or 80 years, for example, a human clone might live only as long as the remaining life expectancy of the adult from which the original donor cell was taken. The reasoning here concerns the role of *telomeres* (the end sequences that cap chromosomes) in cell division – they become shorter with each division, and hence theoretically place a finite limit, determined by their original length, on the life of the organism. There was some evidence from the Dolly experiment that the telomeres were shorter than usual for an animal her age – according to the theory, this would result in a shortened life expectancy.

There are obvious counter-arguments to this hypothesis. More than 90 per cent of all the cell divisions that occur in an animal's life occur in the womb, so it might be expected that the ageing effect in cloning would be minimal. There are other problems with this thesis as well; eggs are packed with enzymes that lengthen telomeres, for example, and species with long telomeres do not live longer than species that have been observed to have shorter telomeres. Finally, researchers have found that telomeres do not always shorten significantly with age. It has been pointed out that the telomere hypothesis may be no more logical than the proposition that since everyone who grows old becomes wrinkled, it is wrinkles that cause old age.

One irrefutable fact is that the cloning of mammals up to now has been a wildly inefficient process. Hundreds of attempts may be necessary both to create an embryo and to implant it successfully. Only a very small percentage of these procedures has been successful in what has been essentially a numbers game.

There is also mounting evidence that cloning can induce degenerative disease (such as the arthritis evident in Dolly), or produce horrific genetic defects. Leading scientists are saying that the tiny minority of animals that survive this traumatic process beyond the embryonic stage – perhaps one or two per cent – often develop life-threatening diseases such as grotesque obesity, developmental difficulties, heart defects, lung problems and malfunctioning immune systems.

The balance sheet

In spite of all this, the initial public revulsion at the notion of cloning – what some bioethicists call the yuck factor – has dwindled as more mammals have been cloned. Even legislative efforts to ban cloning for research and medical purposes appear to have stalled in many countries. Only a few states in the US, for example, have passed laws against it – and the US federal moratorium on such research merely precludes government money from going to it.

Meanwhile an increasing number of bioethicists have begun to accept the idea of cloning technology in view of its obvious potential benefits in medical science. Several scientific groups have already received permission to clone human embryos to the early embryonic stages for the purposes of medical research.

And so the debate continues – a debate that may have momentous consequences, and one that requires the input of a well-informed public.

CHAPTER 5

DNA AND GENETIC ENGINEERING

Genetic engineering has enabled the modification of the naturally occurring DNA of animals and plants, and the commercial production of many biological products that were not previously possible. Among these products are genetically modified food, and many variations of naturally occurring species.

Several of these possibilities raise contentious issues. For example, genetically modified food offers many potential benefits – these include increasing and varying the world's food supply, and using plants to produce vaccines. At the same time, there are fears that the production of genetically modified food may have unintended consequences, and perhaps harm human health in unforeseen ways. Likewise, the genetic engineering of animals – and humans – raises the prospect of benefits such as the use of animals for producing valuable proteins or transplantable organs, alongside more debatable outcomes such as designer babies.

We have already mentioned the procedures used to sequence the component genes of an organism and determine their function. Scientists have also developed the means to cut out selected DNA sequences, and to transfer pieces of DNA between individuals. When you remember that the genetic code is common to all organisms, and consider what we have already said about cloning, you can see some of the dramatic possibilities available to the genetic engineer.

Genetic engineering has enabled the commercial production of new biological products, many of which could never have been derived by previously available techniques.

Applications include:

- increased food production through improved crops and farm animals
- gene therapy
- the biological management of industrial wastes
- the manufacture of new vaccines.

One result of all this is a tremendous increase in the number of new molecules jostling their way through the clinical research pipeline for a place in the biomedical marketplace, including hormones, growth factors, anti-viral and anti-tumour agents, clotting factors, and many new and smart drugs.

Before considering further the extraordinary range of possibilities opened up by the advent of genetic engineering, though, we will have a look at the technologies used to cut and transfer sections of DNA.

Gene technology

As you know, a DNA molecule consists of long sequences of nucleotides whose active units are genes. In round numbers, human DNA contains about thirty thousand genes, and each gene contains in the order of a thousand nucleotide pairs. The genes in the DNA can be compared to the individual negatives in a long strip of film – any particular negative can be cut from the sequence and glued back in a different position, or even into a different strip of film. Similarly, individual genes may be excised from the main DNA strand and either replaced in a different position on that strand, or inserted into a completely different strand, by genetic engineering.

Scientists are now able to use *hydrolytic enzymes* (which simply means enzymes with the ability to split DNA molecules) to release a specific gene from the strand. The enzymes *hydrolyse* the bonds between the nucleotides at each end of the gene, releasing it for further manipulation.

Positioning the gene in the cells of another organism requires more technical finesse. As we have seen, every cell in an animal's body is surrounded by a membrane; this membrane forms a fatty envelope that contains the liquid components of the cell and its *organelles*. At the same time, the membrane resists the entry of water and of substances like DNA that readily dissolve in water. Scientists need to resort to trickery to enable any DNA – such as the gene to be inserted – to pass through the cell membrane and combine with the nuclear DNA within.

One trick is to attach *calcium phosphate* to the gene. The membrane then accepts it as a salt granule, and it is transported across the membrane in disguise. Another is to shock the cell with an electric current, which briefly forces open the pores of the membrane and allows the gene to enter. A third procedure is to bind the gene in a fatty envelope called a *liposome*, which can slide into the cell. A fourth procedure, already mentioned, is to incorporate the gene into a suitable virus which can enter the cell and carry the gene into the host's DNA.

Using one or other of these procedures, scientists can isolate genes from one organism and transfer them to another. This process, called *transgenics*, has revolutionised many areas of biology.

Transgenics

We have already mentioned the possibility of transferring human genes to micro-organisms, using as an example the transfer of the gene for insulin into suitable bacteria. This procedure enables the large scale and rapid production of human insulin – a major improvement on previously available procedures, and of obvious benefit to the more than one per cent of humanity who suffer from diabetes mellitus. This type of gene transfer can be used for the scores of human genes that code for products used in the treatment of human disease, and it represents a major advance in medical biology.

But transgenics is not restricted to human–bacterial transfers. Because the genetic code is uniform across all biological organisms, a multitude of interspecies transfers is possible – including transfers between plants and animals. Here we will consider only a few of the major possibilities.

Many improvements can, potentially, be made in plants by the transfer of genes from other species; once this has been done, copies of the improved plant can be readily produced by cloning. Some of these improvements involve practical features such as an increase in the quality and rate of production of timber; but undoubtedly the most highly debated alterations involve *genetically modified foods*. Similarly, much of the driving force behind the research that produced Dolly and other mammalian clones came from recognition of the benefits to animal breeding that could be derived from gene transfer between mammals. These issues are discussed in more detail later in the chapter.

In this chapter we will also consider gene transfer between human beings.

Plant transgenics

The transfer of genes within and between plant species by genetic engineering can be used to accelerate the changes and improvements obtainable using standard plant breeding techniques. The new varieties can, in most cases, be readily propagated, since the cloning of plants is usually a simple matter. The new methods have been widely implicated in the dramatic results of the worldwide 'green revolution', and in increased plant resistance to factors such as viruses, frost, insect pests and herbicides.

Even more interesting are the possibilities of transferring DNA to plants from non-plant species. A number of developments in this area, such as the production of vaccines from plants, have considerable potential.

VACCINES FROM PLANTS

Genetically engineered plants offer the promise of food-based vaccines – a possibility that should appeal to any parent who has taken a small child for an inoculation. Infections that enter the body through the stomach or alimentary canal are particularly attractive targets for edible vaccines, because vaccines introduced in this way can activate the mucosal immune system, which attacks germs as they invade mucosal tissues such as those of the mouth, nose and gut. Conventional injected vaccines stimulate a systemic immune response, but have little effect on the mucosal system.

One example of such a plant product is a potato that has been engineered to provide immunity against the *E.coli* bacteria that causes travellers' diarrhoea. Food-based vaccines against hepatitis B, cholera, rabies and measles are also being developed.

It is far cheaper to produce vaccines in vegetables and other plants than through the pharmaceutical processing used to make conventional vaccines. Although not every vaccine is suitable for growing in plants for oral ingestion, and the use of plant vaccines is likely to complement rather than replace injections, the value and advantages of these developments are clearly of major importance.

Genes have also been introduced into plants from other species, including non-plant species, to improve their normal performance – to make them, for example, more resistant to extreme changes of temperature, rainfall, salinity, herbicides and insects. This brings us to genetically modified foods.

Genetically modified foods

The case in favour

The world's food supply could, presumably, be dramatically increased by growing genetically engineered crops that can cope with frosts, weeds, pests and drought. For example, weed infestations can very easily be removed without damaging the crop if it has been engineered to be herbicide resistant. Plants have also been engineered to resist insects and viral infections. Indeed, some people claim that this technology offers the only rational hope of feeding a rapidly growing world population.

Supporters of genetically modified foods argue that genetic modifications can improve nutritional quality, availability and flavour of foods, reduce the need for pesticides, and have positive environmental and health impacts – that they are, in fact, just an extension of the processes of evolution and cross-breeding, but much faster and better targeted.

Many of these procedures appear to offer definite advantages. To take one example, scientists have isolated the gene that controls flowering in plants, and manipulated it to make plants bloom at particular times. Modifications of this type could enable farmers to predict to a day when their wheat or canola crops will be ripe for harvesting. They would be able to plant with more certainty of success, taking advantage of the best weather or the best prices; and the floral industry could plant flowers to bloom on cue for events like Mother's Day.

To take another example, one of the more widely grown genetically modified crops is corn. In this case, the DNA of the genetically modified product is generally identical to that of the natural corn, except for an added transgene package (from a bacterium) that enables the modified corn to repel one of its worst natural pests – the bollworm.

Genetic modification enthusiasts say that this technology has been a great success in that it has reduced the amount of pesticide that farmers have to use, and that these products have satisfied all the dietary tests required by US heath regulators.

The case against

On the other hand, many concerns have been expressed about genetically modified foods, and they need to be addressed.

One concern is that consumers of genetically modified foods who are unaware of the nature of the modification may suffer an allergic reaction to, for example, an atypical protein that has been introduced into the food. Critics argue that genetically modified foods have appeared in supermarkets untested, unassessed and unlabelled, and they have put forward a strong argument for the full labeling of such foods and for further testing.

Another concern is that the transgenetic modification of plants is tantamount to playing ecological roulette. By putting the antifreeze genes of the flounder into tomatoes, for example, or chicken genes into potatoes to increase disease resistance, scientists are introducing radical new elements into the balance of nature, with outcomes that may be unpredictable.

It is argued that genetically modified foods ('Frankenfoods' to some) may harm human health in unforeseen ways, and that cutting and splicing genes in a test tube cannot show how they and the resultant organism will behave outside the laboratory. Although a foreign gene inserted into a plant may be innocuous in theory, in practice it may cause a cellular imbalance that results in the generation of harmful products. Some scientific evidence to this effect has already been published in leading medical journals.

There is also the possibility that herbicide-resistant or pest-resistant transgenic plants may lead to the development of new resistant strains of superweeds and superbugs, or that the higher levels of chemical residues from an increased use of weed-killing herbicides may have harmful effects.

What is more, recent research has shown that some of the genes used to genetically modify crops may jump the species barrier. A gene used to modify canola has been found to have transferred to bacteria living inside bees. Such findings appear to contradict claims by supporters of genetically modified foods that the introduced genes cannot spread, and they increase the pressure on farmers and governments to improve existing measures to minimise inadvertent gene transfer.

There are also concerns about the degree of control exercised over genetically modified food production by large multinational companies. It has been estimated that more than half the soya beans and one third of the corn planted in the US comes from genetically modified seeds produced by multinationals with patent rights on genes, processes, and products; and great concern has been expressed that some of these companies are involved in research into *terminator genes* – genes that would render infertile seeds produced from a specific genetically modified crop, so that farmers would not be able to produce their own seeds but would have to purchase new seed every year from the company.

And already farmers have been taken to court and fined for using patented crops without permission, even where the farmer did not plant the crops, which grew through cross-fertilization from neighbours' fields. There is even concern that the control of agriculture from seed to supermarket may soon be monopolised worldwide by a few multinationals.

There is undoubtedly some substance in each of these concerns, and it is up to each society to consider necessary safeguards and controls, and to assess the balance of benefit and risk. This is already an urgent matter; there are millions of hectares of land devoted to the production of genetically modified foods. Concerned citizens need to inform themselves about the issues, and participate in the development of effective biosafety protocols.

Genetic engineering of animals

A major thrust of the research that led to mammalian cloning was the possibility of improvements in traditional animal breeding techniques. Dolly, for example, was produced in an animal research institute, and the ultimate objective was to accelerate traditional methods by selecting an animal for desired traits and producing multiple copies of that animal without having to wait for several generations.

Genetic engineering opens up much more extensive options. Genes may theoretically be introduced into animals not only from the same species, but also from quite different types of organisms (including plants and invertebrates) before the animal is cloned. The biological and ethical implications are even more complex than in the case of genetically modified food.

One possibility that was pursued in the experiments preceding Dolly involved injecting the gene for human insulin into sheep embryo cells, with the object of arranging matters so that the gene would be turned on when the sheep produced milk. This type of procedure would allow animals to be used as 'factories' for the production of foreign proteins.

Many other combinations and permutations of genetic traits are possible. There is even the possibility of producing *chimeras* – animals with characteristics of more than one species, created by mixing embryo cells – and animals such as sheep–goats and sheep–cows have been produced already.

Mind-bending as these prospects are, however, there may be a downside. For example, some of the research on using pig organs for transplant has been halted by fears of disease. The cloning pioneers at the Roslin Institute (where Dolly was born) halted a multimillion-dollar project after scientists in London revealed that pig viruses could easily be transmitted to humans following a transplant, and might spark a lethal epidemic. The Roslin scientists had been working on techniques to rid pigs of specific genes so that human immune systems would not reject their organs. Then came alarming reports on the transmission of cancer viruses between species in the wild. It had been found that a leukemia virus closely related to a virus known to infect pigs was jumping between species. The risk was considered too great, and the project was closed down.

Genetic engineering of humans

Genetic engineering may soon enable humanity to take control of life on this planet in ways that were previously unimaginable. There is an extraordinary potential to improve the quantity and quality of the food we eat, to cure genetic disease, to replace damaged tissues and organs, and to control the processes of growth.

This awesome potential has led many people to consider how we can improve the health and well-being of the developing child, and even how we might manipulate the genes of unborn children for their benefit. The concept of 'designer babies' is one of the most topical – and controversial – in genetic engineering.

In the future, improvements in our understanding of development on a molecular level may make it possible to manipulate the genes of an unborn child to influence such factors as intelligence, athleticism, artistic ability, temperament, even eye and hair colour. These characteristics are often sought for offspring by natural means, of course, through a potential parent's choice of partner; but while the natural process receives universal approval, the possibility of producing such attributes in humans by cloning or gene transfer has not been widely accepted. The possibility exists, nevertheless, and it has been emphasised by recent advances in cloning technology, and in our knowledge of the human genome.

As we have seen, the first rough draft of the three thousand million base pairs that make up the human genome has been released, and full details will be available in the near future (see chapter 4). It seemed to many scientists that these developments would enable them to control the deepest intricacies of the life process. But this was to underestimate the complexity of molecular function, the detail of how human bodies work, and the differences between animals and humans.

The limits of genetic knowledge

The first point to make is that even a knowledge of the base sequence of the human genome leaves our knowledge of life processes in its infancy. It has been estimated that it will take several decades at least to work out the full meaning of this genetic information and the details of its relationships to cellular interactions and the molecular processes of tissue differentiation.

Our knowledge of gene interactions, for example, clearly has a long way to go; and in the human realm, there is a danger that an enthusiasm for genetics may lead us to intervene before we know enough to do it with impunity. Curing one disease may cause others as yet unknown, for example. And there is the possibility of *germ line therapy* – the alteration of genes in the reproductive cells – which would pass any alteration down the generations. This is banned in most countries because its ultimate effects cannot be anticipated.

A second point is that genetics may offer insights into more than just disease states; it may also offer insights into behaviour and personality. A gene that seems to predispose towards violent criminality has been found, for example. What should we do with such information? Would governments require that foetuses carrying a 'criminality gene' be aborted? If they didn't, would they be blamed if that person grew up to commit violent crimes? Would parents be blamed? Is a violent criminal with this gene less guilty than one without it? Our entire system of justice – based on the notion of individual responsibility – may be called into question.

Then there is the question of identity – our culture has always assumed that our biological existence is stable; now, perhaps, it need not be. What about nature? Do we treat it as something that can be infinitely altered to suit humankind, or as an object of respect that we meddle with at our peril? And what if our power over the genome leads to a new free market eugenics in which parents choose to suppress genes for traits they perceive as undesirable? The last time eugenics was tried, tens of thousands of women were forcibly sterilized, and the Jewish populations of some European countries were all but exterminated. Should people be free to abort embryos with 'undesirable' genetic traits? And who decides which traits are undesirable?

Genetic materialism

More subtly, genetics may change our view of ourselves. People used to say 'It's in my stars' – now they tend to say 'It's in my genes'. There are significant dangers arising from such a belief. For one thing, it just isn't true. Whatever else we are, we are more than the sum of our genes. Environmental influences clearly play a part in human development, and the human body will always be more than just chemistry and physics.

Nevertheless, a belief in genetic materialism has its temptations. Many of us, for example, would be interested in the possibility of locating and manipulating a gene that would let us produce bright babies. After all, intelligence is widely regarded as the most prized human attribute, and the one which most clearly distinguishes us from other animals.

But it appears to be a fact that intelligence and other desirable attributes of this type are not produced solely by particular genes. Much of the development of the brain appears to be *epigenetic* – that is, as well as the inherited component, there is an unquantifiable effect arising from the environment and the fiendishly complex interactions of cellular metabolism during development. It happens within an environment created by the genes, but not under their total control. It seems conceptually dubious, therefore, to link specific DNA sequences directly to the quality of output of the developed human brain.

It is also worth pointing out that the qualities that make us human go beyond intelligence alone, and include complex attributes such as creativity, compassion, sensitivity and wisdom.

The concept of human cloning raises additional questions. Is it OK for people to produce clones of themselves, if the technology becomes available? Have we unlimited freedom to tailor the world to our own design?

The questions raised here have an increasing urgency with the appearance on the scene of designer babies.

The birth of the world's first true designer baby was reported in 2000. The baby was genetically selected – although not genetically engineered – to have the type of cells needed to save the life of his dying sister. His birth triggered an intellectual and moral storm.

Although the parents were able to conceive naturally, doctors used the couple's sperm and eggs to create fifteen embryos in the laboratory using IVF techniques. At two days old, one cell from each embryo was removed and DNA analysis was carried out. Embryos found to carry the same genetic disease as the ill sister (Fanconi's anaemia), were discarded – as were those which would not produce a baby ideally matched to her tissue type. Then two suitable embryos were identified, and one was implanted into the mother. After the embryo had developed into a foetus and the baby was born, doctors collected blood cells from his umbilical cord in a painless procedure and transplanted them into his sister. They anticipated that these cells, with their potential to develop into any type of body tissue (see page 18.) would become healthy bone-marrow cells and give the girl a new immune system. For the first time a screened embryo had been implanted into a mother's womb for the purpose of saving the life of a sibling, although laws surrounding the use of IVF technology would prevent this happening again in many parts of the world. >

> This case was judged harshly by some bioethicists, who saw it as playing God and only a small step away from producing babies for the sole purpose of allowing organ donation; but other people took the view that the key issue was that the boy was wanted and loved for himself, not just as a source of spare parts for his sister – in fact, teaching the world that some children with genetic diseases may not have to go on suffering throughout their lives.

CHAPTER 6

DNA AND TISSUE ENGINEERING

The possibility of 'spare parts' for humans with diseased organs has now become very real, with some engineered products already on the market. There are also encouraging prospects of regenerating complex organs from cell samples, and the use of stem cells shows great promise in this direction.

At the early stages of development, individual cells in the embryo (embryonic stem cells) have the ability to develop into any of the hundreds of cell types in the adult organism.

Stem cells have now been isolated from several adult tissues as well, and transformed into different cell types. Scientists are expressing increasing confidence in their ability to program stem cells to replace diseased or damaged tissue.

Arguments for cloning humans in their entirety have not been well received on the world scene. The use of techniques to produce individual tissues, however, has proved far more acceptable. There is a continuing strong demand for 'spare parts' to treat patients with diseased or damaged organs, and the possibility of supplying these through tissue engineering has many attractions.

There is already at least one tissue-engineered product on the market – skin. Techniques had previously been devised for growing sheets of epidermal (skin) cells for treating burns patients, essentially by culturing the patient's skin cells under conditions that favoured their growth and reproduction. Building on this achievement, a product that contained the main cellular constituents of skin (the *epidermis* or outer layer, and the *dermis* or inner layer) was developed, becoming the first biomedical device containing living human cells to be approved for use by the US Food and Drug Administration. Methods of preparing cell suspensions from a patient's skin and then spraying this suspension over affected areas have also been developed, and have achieved impressive results.

Such developments require both a suitable source of cells and an intimate knowledge of the growth requirements for those cells. These factors are well defined for skin, and for many single cell types; but most organs contain several different cell types, arranged in specific ways. The challenge here is to define the *developmental pathways* – the processes by which these organs grow and develop – in enough detail to enable organ-specific requirements to be met. Tissue engineering to meet this objective is theoretically possible, and rapid progress in this direction is being made. We will look now at the nature of these developments, the promise they hold, and their implications for medicine and society.

Strategic objectives

There are two major strategic objectives in the attempt to create organs and tissues suitable as replacement parts for humans:

- injecting a mixture containing various cells, chemicals and growth factors into a tissue requiring regeneration, which would cause the patient's cells to migrate into the appropriate positions and restore the tissue to its normal structure and function
- incorporating the cells (either the patient's own or those of a donor) into a three-dimensional scaffold made of *biodegradable polymers* – large synthetic molecules that are capable of breaking down in the body. After the cells have reorganised to form new tissue the polymers break down, leaving a completely natural tissue. For example, using a biodegradable polymer mould to guide the structural aspects of growth has already enabled cartilage, which has low nutrient needs and does not require the growth of new blood vessels, to be grown in the shapes of ears and noses.

Although both these objectives are ambitious, particularly in relation to the growth of complex organs such as hearts, kidneys or livers, the scientific knowledge necessary for their realisation in the near future is accumulating rapidly. This knowledge concerns the nature of the molecular factors that drive development and differentiation.

A good deal is already known about this key subject; and research is making rapid strides towards a fuller understanding. A number of specific *growth factors* – the proteins necessary for growth and differentiation – have been identified, along with the gene sequences coding for many of them. One example is the *bone morphogenetic proteins* that direct the regeneration of new bony tissue; but there are many other examples from this critical area of research. A sufficient understanding of developmental pathways in the human body will eventually provide tissue engineering with a tremendous range of possibilities.

While tissue engineering through the application of growth factors is a promising development, however, it doesn't meet the ultimate objective of the tissue engineer, which is the growth of whole organs from an initial cell sample. The biochemical and structural factors involved are extremely complex and thus difficult to replicate artificially; and the involvement of the body's natural directive processes appears to be essential.

Cellular engineering

Even if the time has not yet arrived when humanity can benefit from the full potential of tissue engineering, there are already a number of applications for cellular engineering that involve inserting living cells into plastic capsules, which can then interact with the body's metabolism while being shielded from attack by the immune system. Applications in use or in the process of being developed include therapies for serious neuro-degenerative conditions such as Parkinson's and Huntington's diseases, macular degeneration (a common cause of blindness in the elderly), haemophilia, liver disease, and a number of other conditions.

Immunoisolation therapy, as the technique is known, overcomes many of the disadvantages of implanting free cells, which are likely to be destroyed by the immune system unless they come from the recipients themselves (or an identical twin). Cells implanted in a plastic casing also have the advantage that they can be easily retrieved if necessary, whereas free cells often cannot be recovered. The capsules have a membrane pore size that will block invasion by immune molecules, but is large enough to allow the inflow of nutrients and oxygen and the outflow of the proteins released by the implanted cells.

Obtaining sufficient quantities of human cells for the large number of potential recipients in such cases may be difficult, so *cell lines* – cells readily grown in the laboratory – are often used. This opens the door to the use of genetically modified cells – which means that immunoisolation technology has suddenly offered a new way to provide gene therapy. Molecular biologists can insert genes for medically useful proteins into cell lines able to manufacture the proteins, and the cells can then be incorporated into plastic-covered implants. Such devices may function for months, or years.

Much challenging work remains in developing such products to the stage where they have widespread application in medicine, but at the present rate of progress it may not be many years before equipping patients with tissue-engineered tissues – and organs – is as routine as the installation of coronary bypasses today.

The ultimate application of the body's processes in directing differentiation would be the regeneration of complex organs from a cell sample; and as futuristic as this may appear at first sight, there are exciting and encouraging prospects emerging from studies with *stem cells* – undifferentiated cells, mostly occurring in the embryo. These cell types have attracted a great deal of attention in recent times; they are, in fact, becoming a seriously hot biomedical topic.

Stem cells

Stem cell technology has received a great deal of journalistic attention recently, both because of its very obvious potential for public benefit and because some of its aspects have been the subject of much controversy.

Some commentators have prophesied a revolution in the practice of medicine through the amazing capacity of stem cells to organise and direct the processes of growth and tissue differentiation.

As we have seen, differentiation is an extraordinarily complex process; but it is central to an understanding of the directions of much modern biotechnology. While many of the directive factors behind the development of some cell types are now understood, the complexity of the overall processes means that their duplication in the laboratory is technically difficult. Indeed, in many cases, it is impossible at present.

One way of outflanking these technical difficulties is to make use of nature's own competence in these matters, a competence demonstrated over thousands of years of human development. Making use of the natural characteristics of stem cells in this way is one of the most promising areas of biotechnology.

Embryonic stem cells

You will recall that in the initial stages of development the fertilized egg duplicates its DNA, then divides repeatedly to form multiple copies of itself. After a few days, it exists as a hollow ball (the *blastocyst*) containing undifferentiated cells. These cells are the embryonic stem cells, which have the amazing potential to eventually become any of the two hundred or so specialised cell types in the adult human body (that is, they are *pluripotent*).

It is only after the blastocyst implants itself in the uterus after about seven days of development that the stem cells become disparate in form and irreversibly dedicated to special tasks, finally differentiating into blood, bone, muscle, nerve cells and so on. These transformations result from the combination of many natural growth factors, nuclear receptors, complex signalling systems and interactions with DNA.

Adult stem cells

Besides the embryonic stem cells that act as precursors for different cell types in the transition from embryo to adult, there are also stem cells in mature human tissues. These so-called adult stem cells may also be pluripotent, although it is not yet clear whether this is in fact the case. Certainly, they can give rise to different sorts of blood cells, skin cells and gut cells, for example, and continue to give birth to their particular progeny throughout our lives. They live deep in the tissue strata of our bodies, and their function appears to be to help the body repair itself.

Research with stem cells

Research into the control of the developmental processes of stem cells has reached an exciting stage. The ultimate goal is to create cures for any disease that involves living cells, and the results could be as momentous for medicine as the discovery of antibiotics or the initiation of organ transplants. Scientists are expressing ever increasing confidence in their ability to program stem cells; and there is growing evidence that the process could even be reversed, with adult (differentiated) cells being deprogrammed back to stem cells, which could then be converted into whatever type of cell was required. Such a process – or the use of stem cells from adult tissues – would of course dispel ethical concerns about destroying embryos to obtain stem cells.

Directing differentiation

The box on page 93 suggests how new body parts might be grown. Some of the steps are hypothetical, and scientists are still finding out how to manipulate the enormous array of growth factors needed to make the different cell types diversify in a controlled manner. The point is that embryonic stem cells can give rise to all the cell types that are present in the human body, as illustrated in figure 6.1.

The hope is that this flexibility can be managed for the repair or replacement of all damaged tissues; and that hope seems to be realistic in the light of present progress. Surely the developmental code will ultimately be discovered, just as the genetic code has been. Many would regard this goal as the holy grail of developmental biology.

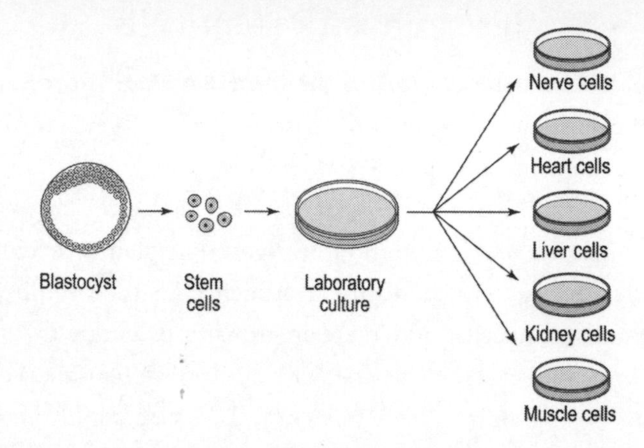

FIGURE 6.1 The potential of stem cell technology. Stem cells are isolated from the early embryo (blastocyst), and cultured in vitro with the appropriate growth factors to allow them to develop into particular cell types such as nerve, heart or liver.

In one example of the progress that has been made in directing the development of stem cells, it has been shown with experimental animals that treating stem cells with *retinoic acid* (a vitamin A derivative) causes them to grow into nerve cells. Apparently this single chemical has the effect of activating a set of genes used only for nerve cell development, while inhibiting the genes involved in the development of other cell types.

Likewise, the complete range of cells normally found in blood have been derived from stem cells by treating them with the appropriate growth factors; and embryonic stem cells have also been coaxed into becoming *hepatic* (liver) and *pancreatic* (insulin producing) cells.

Specialised cells have also been injected into tissues to form stable *grafts*. For example, scientists have allowed stem

GROWING A NEW HEART?

As an example of the possible use of stem cells, consider the theoretical case of a person who has suffered a severe heart attack, resulting in the death of the cells in a major portion of the heart. The prognosis is poor, and the prospects of finding a suitable heart donor even worse. (Around the world there are many more people on waiting lists for transplants than will actually receive them.)

So a small sample of skin cells is taken from the patient, and the DNA from these cells is injected into an enucleated human egg, which is cultured to produce an early embryo. The stem cells from this source (or other sources that we will consider later) are grown under conditions that induce them to develop into heart cells, and are then transplanted into the heart of the patient, where they grow and replace the tissue lost during the heart attack. Because they are an almost perfect genetic match, there are few or none of the usual problems of rejection by the immune system, and the patient returns to normal health and heart function.

cells from mice to partially differentiate, separated out the early heart cells and transplanted them into the adult animals, where they remained viable. Scientists have also been able to grow fully functional human kidneys from embryonic stem cells, using laboratory animals as hosts. Experiments such as these give hope that a number of medically valuable cell types may be developed – skin cells for the treatment of burns or wounds, pancreatic cells for the treatment of diabetes, and cells for regenerating cartilage lost in arthritis, for example.

Growing new nerves

What is more, scientists have recently discovered a way to enable a damaged human brain to grow new nerve cells. Previously it was thought that the brain cells a person was born with were all they would ever have, and that these cells gradually died as the person aged. It has been shown, however, that two parts of the brain – the smell centre and an area for short-term memory – constantly form new brain cells (*neurons*) from adult stem cells. It is now clear that the brain is a highly plastic organ with the ability to undergo rapid changes in neural circuitry at the anatomical, cellular and molecular level.

This is an extremely exciting piece of new knowledge, but the question remains as to how far and fast science can proceed to unlock this potential for the treatment of the many debilitating neurological diseases.

Perhaps, eventually, it will be possible to replace all the differentiated cells of the body. This process may offer a cure for Alzheimer's by enabling the growth of new nerve cells to repair the damage wrought by this devastating disease. It may enable stroke victims to recover. It may offer a cure for people with diabetes, or with degenerative diseases of the liver (such as hepatitis and cirrhosis), the kidneys, the heart or the bones (such as osteoporosis).

THE SUPERMAN SITUATION

A promising start has been made. For example, one of the most tragic cases of tissue damage occurs when the spinal cord is ruptured resulting in paraplegia or quadriplegia, such as that extensively publicised in the case of the actor Christopher Reeves, who played Superman in the movie. In one experiment, olfactory cells were removed from rats, grown in culture, then injected into rats whose spinal cords had been severed. Some weeks later, regeneration of the damaged area was discernable. The rats could walk and climb ladders; and this startling result has led to the first clinical trial of spinal cord regeneration surgery for human paraplegics.

Getting around the immune system

Another challenge for tissue engineering is to produce tissues that are not recognised as foreign by the recipient's immune system. This could be achieved by a combination of cloning technology and the embryonic stem cell technology we discussed earlier.

Cells from the patient (as the cloning donor) could be used to produce an embryo that could be grown in culture until it reached the blastocyst stage and then used to produce embryonic stem cells that were genetically identical to the patient's own cells. This process is no longer theoretical. Researchers have already demonstrated that cloning could work as a source

A DIFFERENT APPROACH

A radically different approach is to allow the embryonic stem cells to differentiate indiscriminately, then find a smart way to pick out the cells that have the desired characteristics. One scientist, for example, has fitted mouse embryonic stem cells with a gene for drug resistance connected to a genetic 'switch' that is thrown only in heart muscle cells. Once the stem cells had differentiated into their usual array of tissues he simply added the relevant drug, which killed off everything but the heart cells. When he injected these cells into mice they integrated into the heart tissue, even beating in synchrony with the mouse's own cells.

of 'grow-your-own' transplants by implanting cloned cells into cattle. These cells formed functioning tissue without any sign of rejection by the animals' immune systems.

Changes in cellular identity

Intriguingly, it is now known that with a suitable switch of growth factors, differentiated cells can be made to regress to embryonic stem cells.

The dogma that once a mammalian cell has taken on its adult task it cannot switch careers is falling apart. The current scientific view is that with appropriate technical intervention, differentiated cells can be deprogrammed back to stem cells, which could then be converted into any type of cell that was required.

The possibility that cells taken from an adult animal can transform from one tissue type to another raises the tantalising prospect of harvesting or creating cells that are just as powerful as embryonic stem cells from an adult body, a prospect with a potentially huge impact. Imagine that a single tube of cells could be turned into enough tissue to treat thousands of patients with different requirements for tissue replacement – an innovation that would leave almost no realm of medicine untouched.

New evidence that cells can switch their identities is rapidly accumulating. Adult neural stem cells have been turned into white blood cells, for example; and it has been shown that when human marrow cells (which normally give rise to muscle and connective tissues) are injected into the brains of rats they behave just like nerve cells, migrating deep into the brain. It has also been shown that skin cells can be reverted to stem cells, then transformed into beating heart cells. The possibilities for self-repair in such transformations are obvious.

Other applications

There are other promising applications of stem cells. Because they can be generated in practically unlimited numbers, they could be useful in research into human proteins that have important functions in the body but are difficult to isolate from tissues because of their very low concentrations. They might also be used to study the possibility that particular drugs might cause birth defects, a major point of scientific and medical interest.

Finally, stem cells offer an approach to studying the early

INVESTIGATING GROWTH FACTORS

Complex signalling systems are needed to guide cells through the elaborate pathways of differentiation to their ultimate tissue destinations, and many of these developmental cues remain a mystery. An increasing number are now known, however. All in all, some hundred different growth factors have been identified – perhaps ten per cent of the total. This is an area where the use of stem cells in the laboratory might be useful in pure research, providing insight into why some cells remain trapped in their developmental fate while others can switch their identities, so allowing a closer molecular definition of the developmental processes.

events in human development in a way that is ethically acceptable. Moral issues that arise in connection with experiments on embryos would not apply if the stem cells were not sourced directly from embryos, and research on these cells could be the means to gain insight into the fundamental molecular mechanisms of human development and differentiation.

Ethical issues

The ethical questions that arise from the use of embryonic material for medical research have proved to be contentious. American laws ban the spending of federal money on human embryo research; in consequence, most pioneering work in the US has so far been sponsored by private companies.

In the UK the situation is now different. An expert panel, backed by the Royal Society, the British Medical Association, the British Heart Foundation and the Medical Research Council, has concluded that the potential benefit of such research outweighs any ethical objections. Many medical scientists regard stem cell research as the most important development in biomedicine for decades, and consider that it would be unethical *not* to proceed when the technology may relieve the suffering of millions. Recently the British government has accepted these arguments, and has permitted the limited cloning of embryos from IVF clinics for this purpose.

In the US, President Bush has recognised both the fundamental importance of stem cell research and the moral dilemma in using human embryos as a source of these cells by making federal funding available only for research on existing embryonic cell lines.

It needs to be made clear that the embryos coming from IVF clinics for research purposes are those that have not been implanted in the mother-to-be; that is, they are unwanted for their original purpose. Many people consider that the use of these embryos for stem cell research is more than justified on the grounds that they would otherwise be destroyed anyway, with no gain to anyone.

In addition, scientists have recently come up with a number of alternatives to *therapeutic cloning*. They have found alternative sources of stem cells such as placentas, the fat of many tissues, and adult tissues such as brain and bone marrow. They can now also dedifferentiate many adult tissues to stem cells. So this is one area of concern that might to be losing its urgency.

The topic of stem cells presents the ethical dilemmas of biotechnology in starker terms than most other areas. The extraordinary potential for medical benefit to humanity has to be balanced against deeply felt ethical concerns.

WHEN IT GOES WRONG

Other ethical questions involve the efficacy, and the propriety, of applying the theoretical potential of stem cells in real medical situations.

Recently, for example, attempts to cure Parkinson's disease by implanting foetal cells into patients' brains produced tragic and irreversible side effects. While the treatment appears to have helped a small number of patients, it was of no benefit to patients over 60 years old, and in 15 per cent of cases the patients ended up with worse symptoms than they had before they took part in the trial. For the first year the patients appeared to progress well, but many then began to develop distressing symptoms, including an uncontrollable jerking of the head, writhing and throwing of the arms. The cells were successfully implanted into the patients' brains, but once there they continued to multiply – in other words, their growth and development were not controlled satisfactorily in their new tissue environment – and this led to an overproduction of *dopamine*, a chemical that plays a vital role in coordinating movement.

CHAPTER 7

DNA AND THE ADULT HUMAN BODY

DNA maintains an active but diminished role in the adult human – at this stage of maturity, its role is mainly concerned with cell repair and replacement to enable the body's tissues to maintain the status quo. At this point, the developmental activity of DNA has culminated in the production of the extraordinarily complicated mechanism that is the adult human body, containing some hundred million million cells and hundreds of different cell types, with maybe a billion molecules of various types in each cell. Maintaining a balance between the functions of all these components is a task of immense complexity.

An understanding of the role of DNA in this period of human life, and its relationship to health and disease, requires some consideration of molecular structures and functions in the adult human body. These functions are far too complex to be dealt with here in any detail; but some of the body's major control mechanisms at the cellular and tissue level are outlined in this chapter.

This basic explanation of metabolic relationships not only puts the role of DNA in the adult into context, but also gives an indication of the present and future directions of biomedical science. The human condition is increasingly being explained in terms of cell biology, and disease and dysfunction in humans will be specified more and more often in such terms.

The initial stages of development and differentiation in the human body occupy about a quarter of the normal total lifespan. The body then reaches a biological plateau, where it usually remains until the ravages of age take their gradual and seemingly inevitable toll.

In the adult human, DNA switches to a role that is mainly directed towards maintenance of the status quo. Cells and their constituents in the adult body have a finite lifetime, there is a continual need for replacement, and DNA plays a central role in this process.

The role of DNA in the adult

What does DNA do in the mature human body? In previous chapters, we have followed the developmental processes during growth and differentiation from a single original cell to an adult stage characterised by the presence of some hundred million million (100,000,000,000,000) cells, and hundreds of cell types. Clearly, DNA replication has been extremely active during this period.

The need for this degree of active replication is greatly reduced in the adult. Even in maturity, however, many of the body's cells and tissues continue to use DNA in the synthesis

MAINTAINING THE STATUS QUO

In the adult human body, millions of cells die and require replacement every second.

Consider the blood supply. Red blood cells, like other cells in the body, have a finite lifespan before they disintegrate and have to be replaced – about 120 days on average – so to maintain normal bloodstream function the entire complement of red blood cells and their constituents (mainly haemoglobin) must be resynthesised and replaced over this period. Other cells have different component molecules and different lifespans; but their worn-out components must also continually be replaced.

In the case of reproduction, the body must replace lost sperm cells, or provide the cellular material required by the growing embryo.

of new cellular material. The cells in adult tissues are not entirely quiescent; there is a continual need for replacement and, of course, reproduction as the DNA is passed on to a new generation.

So even when normal growth and differentiation have come to an end, DNA continues to have an active role in the body's synthetic processes.

The complex transitions in molecular composition and *morphology* (physical form and structure) that take place at a cellular level as a person matures also involve tightly coordinated changes in the *expression* of DNA. Not surprisingly, far more genes are expressed during early development than in other parts of the life cycle.

The rate at which individual genes are expressed at different stages of development – *differential gene function* – is regulated by a number of processes that affect the rate at which their proteins are synthesised and degraded. The most complex regulation of gene expression occurs during the development of the embryo, when the sets of regulatory genes operate in a complex order in space and time, and in a highly coordinated manner. Differential gene expression is under the exquisitely delicate control of a number of factors too complex to explain in the present context. Here we will just mention two of the main factors:

- *regulatory proteins* that bind to DNA, and
- specific *protein growth factors* that direct gene expression towards the requirements of particular tissues – for example, nerve growth factor plays a critical role in the development of neurons, and epidermal growth factor stimulates the growth of skin cells.

The expression of DNA is the process by which the genetic information it carries is converted into the molecules needed for the structures and functions of a cell.

These forms of regulation, so intimately involved in the processes of embryonic development and change, diminish considerably in the adult. There remains, however, the requirement for *homeostasis* (an equilibrium of function and chemical composition), and the need for complex control systems to maintain a state of balance and health.

DNA and subcellular structure

This section briefly explains the fine structure of a cell. Since cells are the basic building blocks of the body, some knowledge of their structure and function is essential to an understanding of the workings of the body in health and disease.

Some readers may find aspects of the following discussion a little technical, but it introduces terms that you are likely to hear more and more frequently in reports and discussions of medical diagnoses and biotechnology.

As we have seen, in the adult human there are some hundreds of different cell types; but all these cells share certain features. A typical cell structure is shown in figure 7.1.

The cell's DNA carries genetic information, as we have seen. Most of the DNA is tightly packed within the *nucleus*, which is surrounded by a nuclear membrane.

Some idea of how much DNA there is, and how tightly it is packed, can be gained from the fact that if all the cellular DNA in one adult human was extended and put end to end, it would reach from the earth to the moon and back several hundred times.

The boundary of the cell is defined by the *plasma membrane*, which acts as a flexible barrier and contains a number of *receptors* to receive signals from outside the cell, as well as transport systems to allow the passage of certain key chemicals. The *cytoplasm* inside the cell membrane consists of the *cytosol* and *organelles*. The cytosol is a concentrated solution of proteins, RNA (another form of nucleic acid), and organic and inorganic substances, in which the organelles and a number of other

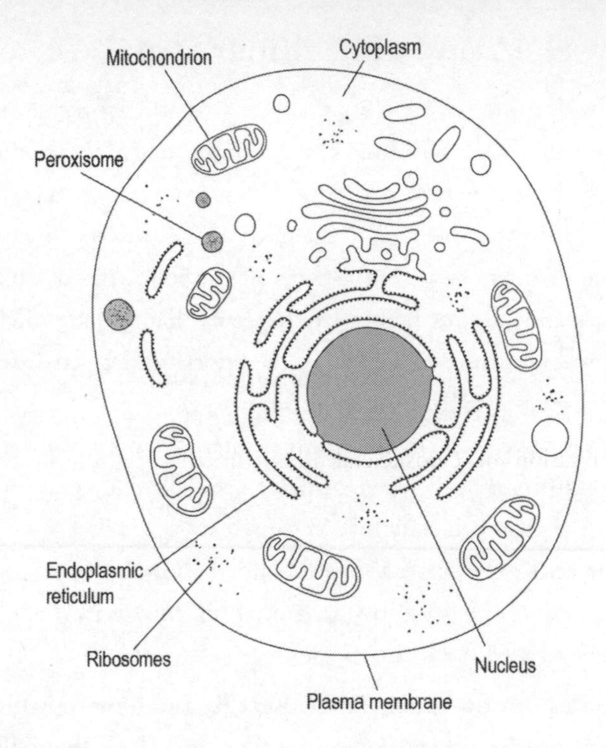

FIGURE 7.1 The major subcellular components of an animal cell. Each part is separated from the rest of the cell by a membrane. The cytoskeleton (a network of fine filaments) is not shown for reasons of clarity.

particles are suspended. Organelles ('little organs') are membrane-bounded structures with specialised functions. The *mitochondria* are organelles that contain some DNA, as explained in chapter 1; they are also major sites of energy generation, as are other organelles called *peroxisomes*. The *ribosomes* contain RNA, and are where protein synthesis takes place.

Eucaryotic cells – that is, cells that contain nuclei, such as human cells – also have a complex system of internal membranes called the *endoplasmic reticulum*. The complexity and extent of these structures in the adult human body is indicated by the fact that the total area of these membranes in a single person may amount to some hundreds of hectares.

The genetic material in eucaryotic cells is organised into *chromosomes*, highly ordered complexes of DNA and proteins. Before cell division, each chromosome is replicated, and the two chromosomes are separated by a process called *mitosis*.

The *cytoskeleton* is a network of various types of filaments and microtubules within the cell; it gives shape to the cell, and it is the reorganisation of cytoskeletal filaments that causes the shape changes characteristic of developing cells. Organelles move along filaments of the cytoskeleton, propelled by proteins such as *myosin*.

In multicellular organisms there is a division of labour between various types of cells; for example, red blood cells are specialised for the transport of oxygen from the lungs to other body tissues, muscle cells are specialised for the performance of mechanical work, brain cells function in the coordination of nervous impulses, and certain pancreatic cells produce insulin to help control blood sugar levels. There are more than two hundred different cell types in the adult human body; each has its own distinctive contribution to make to the normal functional capacity of the individual.

DNA and metabolic controls

By any standards the human body is an extremely complex structure, and it is only able to continue in a normal healthy state due to the presence of an extraordinarily intricate series of controlling checks and balances.

We have already mentioned the regulation of differential gene function during human development – now we are superimposing metabolic controls at the level of individual cells and tissues. There are many layers of regulation at this level, too – complex, interrelated systems that are beyond the scope of this book to describe in detail. Major control processes at the cellular level include:

- nucleic acid metabolism (that is, the processes involving the synthesis and breakdown of nucleic acids)
- gene expression (which we have briefly considered)
- regulation of the rate of *enzyme* metabolism
- transport systems
- energy generation
- protein interaction.

Protein interaction

Let's look at just one of these processes – how *protein interaction* is involved in metabolic control. Proteins are major functional molecules in cells, and because they are very large, very numerous, and very close together, they tend to bump into one another and cohere with varying degrees of affinity. This might seem to be a relatively trivial phenomenon, but this interaction between proteins may have profound effects on human metabolic processes – a fact that has been fully appreciated only in the last decade or so, with Nobel Prizes in medicine having been awarded for work in this area.

One example of how protein interaction exerts metabolic control involves the relationships of the *glycolytic enzymes* and the cytoskeletal proteins during development. The glycolytic enzymes in cells and tissues are mainly concerned with the breakdown of sugars to provide energy. Sugar metabolism is of particular importance to the developing embryo, with the breakdown of one sugar, *glucose,* being the main source of energy in the early stages. At this point the cytoskeleton has different structural characteristics from those seen in adult tissues, to cope with the special developmental requirements of active cell division, *morphogenesis* (structural development) and cell migration (the movement of cells to their appropriate locations). *Glycolysis* – glucose breakdown – provides most of the energy required by these processes. Alongside the high rate of glucose breakdown characteristic of cells in the foetus, a much higher level of *adsorption* of the glycolytic enzymes to the cytoskeleton than occurs in adult tissues has been observed. (Adsorption just means a process by which a substance forms a film or layer on a solid surface, in this case the cytoskeleton).

Enzymes are proteins that cause chemical reactions in the body's cells to start or speed up.

There is strong evidence that this type of association between enzymes and cell structure plays a significant role in meeting the specialised energy needs of the differentiating cell types during early development, as well as allowing increased flexibility and control of glycolysis in some specialised situations. There is also abundant data showing that the interaction of many other proteins with cellular structure is a dynamic process that is responsive to, and influences, cell metabolism.

Variations in the degree of association between enzymes and cell structure have not only been demonstrated for normal physiological changes; there are also an increasing number of reports of such variations during cellular abnormality and dysfunction.

Tumor cells, for example, display atypical characteristics of both cytoskeletal structure and carbohydrate metabolism. Many scientists compare what goes on in tumor cells to dedifferentiation – the opposite of the cell differentiation process observed during development. The rate of glycolysis in tumors is typically much higher than the rate in normal tissue – just as it is in foetal tissues.

There is increasing evidence that something similar occurs in diabetes. Along with decreased insulin release, and a reduction in interactions between circulating hormones and receptors on the cell surface, there is a reduction in the degree of adsorption of the glycolytic enzymes to the cytoskeleton. This contributes to a reduction in the cellular transport processes for sugars and their utilisation in the body – factors very relevant to diabetes, a disease characterised by disturbances in sugar metabolism.

Tissue metabolism

A very different problem is presented by another of the body's metabolic requirements – control over the interactions between cells and tissues. Assuming that there are adequate control processes within the individual cells of the human body – all one hundred million million of them – how does the body maintain a balance between the separate groups of cells called *tissues* – liver, kidney, brain, heart, pancreas, skeletal muscle, adipose tissue and so on – and their particular metabolic processes?

One of the major ways by which the human body regulates itself at this level of metabolism is through the action of *hormones*.

Hormones

Clearly there is a need for coordination between the various activities of the tissues in different parts of the body; but how is this achieved? One of the main methods used by the body for this purpose is the circulation of hormones in the bloodstream. Since the bloodstream links all the organs, it enables hormonal signals to be carried between them.

Hormones are chemical messengers (which can be *peptides*, *amines* or *steroids*) released by *glands* into the blood. Their job is to regulate the activity of other tissues, and they act in a complicated hierarchy of functions. For example, nerve impulses stimulate the *hypothalamus* (part of the brain) to send specific hormones to the *pituitary* gland, stimulating (or inhibiting) the release of further hormones. These hormones in turn stimulate other glands (such as the *thyroid*, the *adrenal glands* or the *pancreas*) to release their characteristic hormones, which in turn stimulate specific target tissues. Some of these relationships are shown in figure 7.2.

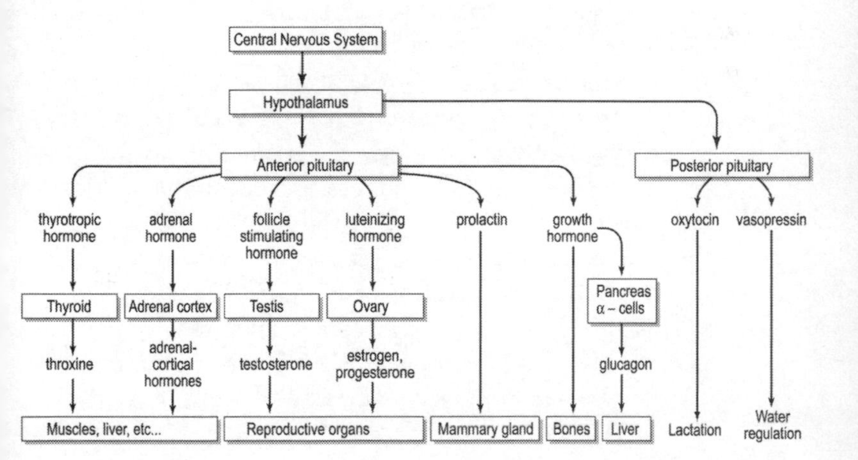

FIGURE 7.2 The interrelationships of the hormonal network in adult humans.

HORMONES AND BLOOD GLUCOSE LEVELS

The concentration of glucose in the blood is hormonally regulated. Fluctuations in blood glucose due to dietary intake or vigorous exercise are counterbalanced by a variety of hormonally triggered changes in the metabolism of several organs. A hormone called *epinephrine* prepares the body for increased activity by mobilising glucose from carbohydrate stores. Low blood glucose results in the release of the hormone *glucagon*, which stimulates the release of glucose from stores in the liver and shifts liver and muscle metabolism to the use of *fatty acids* for fuel, so that available glucose can be used by the brain. High blood glucose elicits the release of *insulin*, which speeds the uptake of glucose by the tissues and promotes the storage of glucose as *glycogen* (mostly in the liver), or fat.

In untreated diabetes, insulin is either not produced or not recognised by the tissues; the utilisation of blood glucose is compromised, and when blood glucose levels are high, glucose is simply excreted into the urine instead of being used by the tissues. The tissues then depend on fatty acids for fuel (producing *ketone bodies*), and cellular proteins are broken down to make glucose. Untreated diabetes is characterised by high glucose levels in the blood and urine, and the production and excretion of ketone bodies.

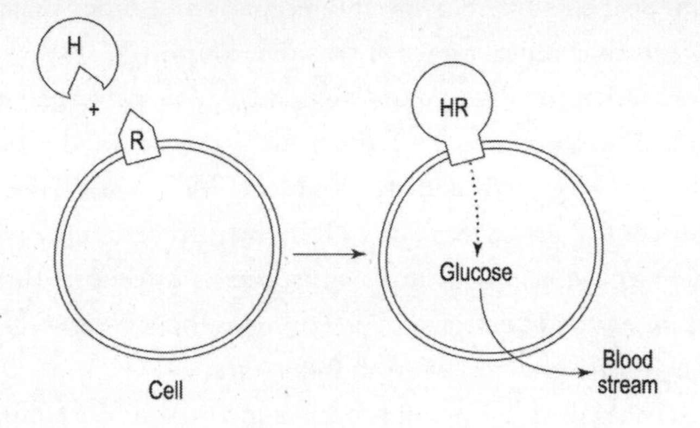

FIGURE 7.3 How a hormone may act at the cellular level. Hormone H interacts with a specific receptor, R, on the surface of a cell, bringing about a series of reactions in the cell; in this case, the release of glucose into the bloodstream.

Figure 7.3 shows the operation of a hormone at the cellular level.

Hormones achieve their specificity and sensitivity through a small number of fundamentally similar mechanisms. Epinephrine, for example, binds to specific receptors on the outer face of *hepatocytes* (liver cells) and *myocytes* (muscle cells). When a receptor is occupied, a series of reactions is activated which causes the glucose stores in the liver and muscles to release glucose into the blood stream. There is a cascade of events in which a single molecule of the hormone activates a *catalyst* (a substance that causes a reaction to start or accelerate) that in turn activates another catalyst and so on, resulting in large *signal amplification*; this is characteristic of all hormonally activated systems. In other mechanisms, hormones use different specific receptors and other signalling systems in complex arrays of sensitive controls.

A variety of diseases are associated with defects in these control systems. Many of them are involved in the proper direction of growth and development. Mutation of the relevant control genes in a cell's DNA may, for example, permit uncontrolled cell division (such as occurs in cancer) through the formation of defective signal proteins that are insensitive to signals from growth factors or hormones.

The array of hormonal processes involved in the control of tissue metabolism in the human body is very wide, and capable of adapting to a broad variety of control requirements.

CHAPTER 8

DNA AND AGEING

The concept of a maintenance role for DNA during the decades of adulthood raises the spectre of impermanence – of a limited biological lifespan for nucleic acids and proteins, implying the natural occurrence of both synthesis and degradation of the body's components. That is in fact what happens; and the balance of these processes changes during the individual's lifetime. During the growth period the metabolism of nucleic acids and proteins is biased towards synthesis; during adulthood, synthesis and degradation are in approximate balance; during ageing, the rate of degradation increases while the rate of synthesis declines, until eventually there is not enough viable structure for the body to maintain a functional existence.

To understand the role of DNA in ageing, we need to consider the nature of these degradative processes.

Human growth and development normally continues until the late teens or early twenties. Up to that time the changes are continuous, overt, and largely programmed by the DNA of the individual.

Note that word 'largely', though. We should not fall into the deterministic trap of crediting all developmental changes to DNA. Nutrition, exercise and general environmental conditions have clearly contributed to the physiological and morphological condition of the body as well at this age. Both nature and nurture are significant in establishing the form of the adult body.

Following the period of growth the role of DNA undergoes a discernible change, as we saw in chapter 7. No longer is it the concert-master, directing marked changes in the form and function of all parts of the human body. It moves into maintenance mode, and is now mostly concerned with repairing the wear and tear on nucleic acid and protein stores in the various organs and tissues, and maintaining a relatively stable physiological status for the body.

Physiological ageing

We are all familiar with the common accompaniments of ageing – as the cells that nourish it atrophy, hair thins, and greying and balding occur; the ageing brain slowly loses tissue in some regions, leading to slowing reactions and faltering memory; the pupil's ability to control light entering the eye diminishes, focusing becomes more difficult, and cataracts and macular degeneration may occur; artery walls thicken and lose their elasticity, raising blood pressure and increasing the risk of heart attack or stroke; bones become porous and brittle; wear and tear on the joints leads to osteoarthritis.

COMMON SIGNS OF AGEING

Hair	As the hair follicles atrophy, the hair on a person's head becomes sparse. The extent of greying and balding is largely controlled by the genes.
Eyes	The eye muscles cannot contract the lens sufficiently to allow focusing on close objects. Cataracts and macular degeneration may also develop.
Arteries	Cholesterol and calcium build up in the artery walls, reducing their elasticity and raising blood pressure, all of which increases the risk of heart attack or stroke.
Joints	Osteoarthritis develops in many older people; joints become stiff and movement is painful.
Bones	With age, there is increasing loss of bone cells and bones become porous and brittle.
Brain	Some regions of the brain slowly lose tissue, or undergo change resulting in slower reactions and diminishing long- and short-term memory.

Virtually every part of the body becomes less efficient with age – probably at a rate of around one to three per cent a year from middle age on, with the rate accelerating. Depressing as this might be to many people, it also serves as a clue that there are widespread biological processes at work.

Can ageing be reversed?

Many theories have been advanced about the degradative changes that are collectively termed ageing, and the ways in which they might be arrested or reversed.

Many of these theories have been based on the idea that it might be possible to artificially extend or imitate the growth processes, and have considered the effect of some of the hormonal substances known to be important in development.

Many hormonal treatments do have short-term benefits, or are useful in the treatment of specific conditions, but no single hormone has been shown to provide long-term anti-ageing effects.

NOT YOUNG AGAIN

In one series of experiments a group of older men were given injections of human growth hormone. There were some temporary improvements (in terms of muscle mass and skin thickness, for example); but there were also unpleasant side effects, including diabetes-like symptoms and inflammation of the nerves. The treatment was also very expensive. When it was stopped, all the ephemeral indications of youthfulness disappeared.

Other hormones, including melatonin and derivatives of the male sex hormones, have also been tried without achieving the aim of restoring youth.

Causations of ageing

Rather than looking for agents that restore youthful appearance and function to the ageing human – 'elixirs of youth' – many scientists have turned their attention to the causes of ageing. Perhaps, if the causes were understood, the process could be slowed down.

In particular, scientists are looking at the role of DNA. In addition to its short-term function in the synthesis of nucleic acids and proteins, DNA now enters the picture as a long-term factor.

First, there is the capacity of DNA to repair itself. We know that the DNA in the nucleus of the cell holds genetic information about the cell in the sequence of its paired nucleotides. Over the course of a life, however, radiation and other processes can damage the structure of the DNA. If one nucleotide of a pair is knocked out, the DNA can easily repair the damage, using the other nucleotide as a template. However, if the damage occurs during cell division, when the helix splits down the middle and exposes single-stranded DNA, it may be irreparable. If a nucleotide is knocked out of this single strand, the DNA may insert a new nucleotide at random, often causing a mutation; and accumulated mutations may affect the synthesis of important functional components in the cell, causing a deterioration of the body's condition. Clearly, DNA repair plays an important part in the control of the ageing process.

Recent DNA research has also shown that genes can influence lifespan. We are all aware that longevity tends to run in families, and is at least partly inherited. Now scientists have discovered genetic mutations that appear to affect ageing specifically.

TELOMERES REVISITED

Another way in which DNA may be implicated in the ageing process involves the role of telomeres in cell division. Before a cell divides, it copies its chromosomes to give each new cell a complete set. In most cells, however, this process does not include the long spirals of DNA called telomeres that protect the ends of the chromosomes. Telomeres get shorter with every cell division. Some scientists believe that they eventually become so short that the cell can make no further divisions, and it becomes vulnerable to damage and decay. If this is the case, there may be an inbuilt limit on the human lifespan dictated by the length of the telomere.

We considered this theory in our discussion of cloning in chapter 4. At this stage, the role of telomeres in the ageing process, if any, remains controversial.

People with Werner's syndrome, for example, have a specific gene mutation that appears to accelerate the ageing process. People with this disease become wrinkled and grey in their twenties, often getting cancer or heart disease in their thirties and dying in their forties. The available evidence indicates that this mutation shortens life by reducing the body's ability to repair the damage that is a by-product of normal metabolic processes.

This raises the question of whether perhaps the secret of longevity lies in something as basic as counteracting the damaging side effects of metabolism. Certainly many scientific studies have shown that laboratory animals live as much as 50 per cent longer if their caloric intake is drastically reduced – and it seems that this occurs not because they lose weight (although of course they do), but because there is a corresponding reduction in their metabolic rate.

Oxygen free radicals

We now know that the metabolism of nutrients in the human body – especially the process by which the food we eat chemically combines with the oxygen we breathe to provide the energy necessary for our continued existence – produces substances called *oxygen free radicals* as a by-product. Indeed, about five per cent of the oxygen that humans breathe is converted to oxygen free radicals. These substances are extremely reactive, and can cause damage by combining with essential components in the body in a reaction called *oxidation*.

Usually, most of these free radicals are mopped up by *antioxidants* and normal body processes, but an excess of free radicals can cause extensive damage to DNA, proteins and membranes in the cells and tissues.

The action of oxygen free radicals on cellular structure can be compared to the effect, over time, of rust (which, incidentally, is also the result of an oxidation reaction) on an object made of iron – a general weakening of structure. The action of free radicals on cellular structure may be considered as a sort of biological rust, damaging and weakening the cellular features, and eventually causing diminished physiological abilities.

Oxygen free radicals are extremely reactive by-products of metabolism that can cause extensive damage to the body's cells.

It seems that old cells may be built to a lower standard than young cells. Scientists have studied gene regulation in cells from people of various ages, and discovered that in any individual one per cent of the six thousand genes studied altered dramatically during ageing, increasing or decreasing the level of proteins produced more than twofold.

This suggests a radical change in thinking about ageing. In the past, scientists have believed that ageing was a disease in which cells stop dividing: this study suggests that ageing is really a failure of quality control. As people get older, altered gene function results in cells with diminished function.

It is well-known that diet is important in dealing with the ageing process. The usual advice is to lower the intake of fats, increase consumption of fruit and vegetables and exercise regularly, and it is significant that all these things stimulate the production of antioxidants.

Remember how DNA is distributed in humans cells? Most of it is in the nucleus, but there is a significant amount of DNA outside the nucleus, in the mitochondria. In humans, about 70 per cent of the oxygen consumed in most tissues is thought to be used by the mitochondria. This makes them a major site of free radical formation, which in turn means that the mitochondrial DNA is particularly vulnerable to damage. There is evidence that this results in a cumulative decline in the function and numbers of the mitochondria, which would make the mitochondria particularly significant in many of the degenerative processes associated with ageing.

The involvement of oxygen free radicals in several degenerative diseases associated with ageing is supported by a number of scientific studies. There is a long history of association between free radical reactions and heart disease (atherosclerosis and hypertension), cancer, degeneration of the central nervous system, immune response deficiencies, senile dementia of the Alzheimer's type, and many other conditions. It is now widely accepted by scientists that the changes induced by oxygen free radicals are the source of many, if not all, of the illnesses of ageing, as well as the usual symptoms of 'healthy' ageing.

Defences against free radicals

In most cases, the body has both a repair capacity and defence systems to minimise the effects of free radicals on major sites of damage such as DNA, proteins and membranes. DNA repair mechanisms can cope with minimal damage, but beyond this may lead to interference with normal function, or even degenerative processes such as cancer. Damaged proteins and membranes can also normally be restored or replaced through the body's capacity to synthesise new components; but once again there are limits to this capacity, and it can be exceeded by prolonged damage over time.

The defence systems that limit damage by free radicals may take the form of natural antioxidants such as vitamin C or beta-carotene (vitamin A), or of enzyme systems (such as *catalase*, *glutathione peroxidase* and *superoxide dismutase*) specifically designed to deal with the problem. We'll have more to say about this later.

We have already mentioned that reducing dietary intake, which has the effect of reducing the number of free radicals produced, has been associated with increased longevity in experimental animals. The addition of antioxidants to the diet has also been shown in a number of studies to increase the lifespan of animals, by up to 30 per cent.

DIETARY ANTIOXIDANTS

Oxygen free radicals have the potential to cause extensive damage to our cells and tissues. The body's defences against such damage include enzyme systems that act to neutralise or lessen the harmful effect of the oxygen free radicals. There are also a number of naturally occurring and synthetic antioxidants.

Among the most common natural antioxidants available in our diet are vitamin C, vitamin E and beta-carotene. Vitamin C, a common constituent of some fruits (such as oranges and strawberries) and vegetables (such as cabbage and broccoli) is mainly effective in the aqueous part of the cell (the cytoplasm). Vitamin E (found in vegetable oils and sunflower seeds, for example) and beta-carotene (obtainable from vegetables such as carrots and sweet potatoes) mainly act to protect cell and tissue membranes against free radical damage.

Peroxisomes

We have already encountered some of the subcellular structures called organelles – the mitochondria and the cell nucleus. It is now time to become more closely acquainted with another organelle – the *peroxisome* – which is assuming major significance in studies of growth, development and ageing. Recent data indicate that the peroxisome plays an important role not only in guiding the complex processes of growth and development, but also in countering the potential of oxygen free radicals to cause cellular damage at later stages.

There are also indications of a close connection between peroxisomal function and ageing. Peroxisomes are significantly altered in aged animals, for example, and it has been proposed that these alterations may contribute to the disturbances of fat metabolism and the production of free radicals that appear to be of fundamental importance in the ageing process.

But it is in relation to the removal of oxygen free radicals that the peroxisome may play its most important biological role. There is, as we have seen, accumulating evidence that oxygen free radicals are implicated in many of the disease states associated with ageing, and the peroxisome has come to be recognised as the principal organelle involved in protecting tissues against this type of damage. It does this mainly through its unique and flexible assembly of powerful enzymes, which have the capacity to inactivate oxygen free radicals.

It seems that the peroxisome exerts an appreciable influence on major aspects of the ageing process.

Degenerative disease

Degenerative conditions associated with the ageing process include cancer, cardiovascular disease, rheumatoid arthritis, the degeneration of neural and retinal function, pulmonary emphysema and diabetes mellitus. Damage by oxygen free radicals – *oxidative* damage – has been implicated in each of them.

In cancer, the major modifications caused by free radicals appear to occur in DNA.

There is an enormous ongoing load of damage to the DNA in human cells. Recent research has shown that the number of free radical 'hits' to the DNA in humans is about ten thousand

COMMON DEGENERATIVE DISORDERS OF AGEING

A few of the more common degenerative disorders typically associated with ageing are:

- atherosclerosis
- hypertension
- rheumatoid arthritis
- musculoskeletal disorders
- Alzheimer's disease
- depression
- immunodeficiency disorders
- visual acuity loss
- cancer.

per day, for example. This is of small consequence for the cell until it begins to divide, when cell division converts the damage in DNA to mutations. There are several reasons for this, one of the more significant being that single-stranded DNA is more vulnerable to mutation than the protein-clad double stranded DNA in chromosomes. Also, as we have seen, single-stranded DNA cannot accurately repair itself. Although other factors may be involved, including genetic predisposition and environmental influences, there is much evidence that in most cases the development of cancer may be traced back to oxidative damage to DNA.

There is increasing evidence, too, that many non-cancerous degenerative diseases are the result of unrepaired oxidative damage to the protein or fat components of cells and tissues. The characteristic loss of flexibility in joints that accompanies ageing, for example, may be attributed to oxidative damage to connective tissue proteins. Evidence implicating free radical reactions to lipids (fats) in the disease processes of heart conditions such as atherosclerosis and hypertension has been around for a long time.

Oxygen free radicals have also been implicated in the formation of the *neuritic plaques* in the brain associated with Alzheimer's disease. Brain tissue is particularly sensitive to oxidative damage, and to the degenerative changes in cellular function that occur during ageing. It is relevant that the brain is not particularly rich in protective antioxidant enzymes, and that the chemical 'trapping' of oxygen free radicals has been shown to moderate the damage and leads to significant improvement.

AGEING AND SUGAR–PROTEIN BONDING

The ageing process, of course, is extremely complex; and it must be emphasised here that it involves not only cellular components other than DNA, but also processes other than the action of free radicals.

One example is the tendency of the body's sugars to form tight bonds with proteins over time, eventually interfering with their normal function. The stiff joints that are one of the hallmarks of old age, for example, may be due to sugars combining with the protein *collagen*, as might thickened arteries in older people. Diabetics also tend to age rapidly because of processes of this type.

Sugars are an essential source of energy, but once in circulation they can act as molecular glue, attaching themselves to tissue proteins and cross-linking them into hard yellow-brown compounds known as *advanced glycation end products* or AGEs. Generally, piling on dark pigments in the teeth, bone and skin is harmless. But where glucose forms tight bonds with the long-lived protein collagen the result is a constellation of changes, including thickened arteries, stiff joints, feeble muscles and failing organs – the indications of a frail old age. Researchers now claim to have developed a compound that might rejuvenate hearts and muscles by breaking the stiff sugar–protein bonds that accumulate as we get older.

It seems that free radical activity is of very considerable significance in the ageing process, and that damage to the DNA is a major component in the associated decline in the functional activities of the human body.

Slowing the ageing process

Our knowledge of ageing on the molecular scale has advanced considerably in recent years. As we have seen, DNA is generally implicated at the level of both telomeres and DNA damage. There are also a number of degenerative processes that reduce the functional quality of the body tissues, and chief among these appears to be the damage caused by oxygen free radicals. Telomeres may place an ultimate limit on the degree to which human life may be extended, while increasing control of the processes by which DNA, proteins and other molecules are degraded may well hold the secret to an improved quality of life during old age.

Most of the changes we associate with ageing – greying hair, loss of visual acuity, loss of athleticism, the common occurrence of degenerative disease and so on – may be traced to the degradative processes that become increasingly active with age. Genetic factors may also have a part to play, but it is now widely accepted that lifestyle factors generally have an even greater role in guiding the ageing process.

Clearly, the possibilities of protection against all this biological wear and tear, or of treatments for the 'molecular fatigue' involved in specific conditions, are of widespread interest. It is becoming well known, for example, that antioxidants may impede, or reverse, the cellular damage caused by free radicals. On these grounds it seems that more people should be including these factors in their diet, taking a reasonable degree of exercise, and considering measures designed to prevent rather than alleviate degenerative conditions. The central role of peroxisomes and the enzyme systems involved in removing free radicals also deserve more emphasis in future therapies for these conditions.

The free radical theory of ageing predicts that the healthy lifespan can be increased by minimising deleterious free radical reactions, while not significantly interfering with those essential to the economy of the cells and tissues in the human body. The available data indicate that this may best be achieved by keeping body weight down, and eating a diet that is adequate in essential nutrients but designed to minimise random free radical reactions in the body. A balanced diet of this type would contain only small amounts of components with a tendency to enhance free radical reactions, such as polyunsaturated fats, and larger amounts of substances capable of decreasing free radical damage, such as a variety of antioxidants. It would also include dietary factors such as omega-3 fatty acids that help peroxisomes to proliferate in the body's tissues.

Such an approach might well reduce the incidence or severity of degenerative diseases, and, perhaps, add several years to the span of healthy productive life.

CHAPTER 9

DNA AND ABERRANT DEVELOPMENT

The processes of growth and differentiation in the human embryo and foetus are extremely complicated, and sensitive to delicate balances and interactions. A number of factors can interfere with this process. When things go wrong, the result can be disease, abnormality or even death before delivery.

One measure of the impact of such disturbances of development is the loss of life years that can be attributed to them, which was assessed some years ago as being greater than that caused by all heart disease, cancer and strokes.

We discussed in chapter 2 some of the disorders that can occur when things go wrong during development. Here we will consider aberrant development from a different angle.

Aberrant differentiation has many aspects – too many to deal with fully in this one chapter. Instead of trying to consider them all, we will look at examples that illustrate the variety of significant disorders that may be listed under this heading, in addition to those we have already considered in chapter 2.

Here we will consider some of the dietary and environmental influences that may bring about abnormalities in the embryo and foetus, and some of the aberrant developments associated with cancer.

We will also look at non-natural techniques that can lead to normal development – the techniques of in vitro fertilization.

Aberrant development and pregnancy

It needs be emphasized again that the development process of the human body described in previous pages is one of great complexity and sensitivity. Numerous factors, many of them as yet unidentified, influence the complex and delicate changes in the nature and positioning of the cells that enable development to proceed to the birth of healthy children.

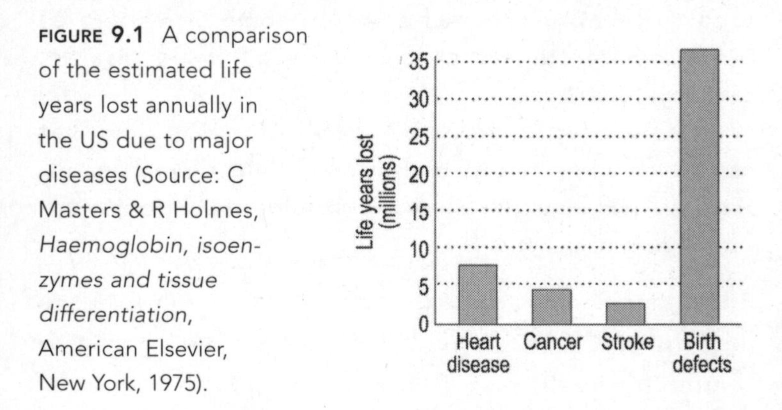

FIGURE 9.1 A comparison of the estimated life years lost annually in the US due to major diseases (Source: C Masters & R Holmes, *Haemoglobin, isoenzymes and tissue differentiation*, American Elsevier, New York, 1975).

FOETAL ALCOHOL SYNDROME

Let's consider the consequences of a heavy intake of alcohol by a mother-to-be during pregnancy.

We know that development and differentiation occur as a result of numerous interactions between key molecules in the embryo and foetus. It's not surprising that this delicate complexity can be readily disturbed by outside agents.

It has been shown that the foetus is very sensitive to maternal alcoholic intake, for example. While there are different opinions about the lower limits of safe intake, there is no doubt at all about the damage caused by a heavy maternal intake of alcohol. Its effect on the developmental processes of the brain is such that the World Health Organisation has nominated *foetal alcohol syndrome* as the leading cause of intellectual disability in children.

Heavy alcohol intake during pregnancy can cause the child to suffer learning disabilities, hyperactivity, and attention and memory deficits, along with an inability to manage anger, difficulties with problem solving, and a poor understanding of cause and effect relationships. The child may be impulsive and easily led. It is not unexpected that a high proportion of young people in detention (more than 50 per cent in some areas) have been diagnosed with this condition – a shocking commentary on one of the more avoidable aberrations of pregnancy.

Recent research has shown that part of the effect >

> of the alcohol is due to damage to certain functional portions of the DNA in the developing organism. The normal operation of several genes is affected – genes that are all associated with the production of *myelin*, the insulating material that surrounds nerve fibres. The reduction in normal myelin synthesis can, not surprisingly, lead to defective brain function.

Up to now, medicine has coped with this complexity in a largely empirical manner. Abnormalities in infants have been traced back, where possible, to their cause, and general advice is given to pregnant women based on this experience – for example, to limit smoking and alcohol intake, and to eat a balanced diet including folic acid.

There is no denying the value of such advice. But if the consequences of harmful influences could be examined and understood in molecular detail, advice to prospective mothers could be improved, and more specific preventative measures could be taken.

The influence of prenatal environment

Research now suggests that there may well be a more general relationship between what happens during pregnancy and a person's health in adulthood.

It is well known that much disease is caused by interaction between the genes received from parents and a person's lifestyle. If a person chooses to smoke, for example, they flirt with cancer; if they choose to eat a lot of fatty foods

and drive everywhere instead of taking opportunities to walk, they may set themselves up for a heart attack. But a new field of research proposes a third factor in the equation: that the environment in the womb can influence the process of foetal development, which in turn influences normal physiological relationships; and that this can lead to the person being at greater risk of diseases such as hypertension, heart disease and diabetes in adulthood.

It seems, for example, that the system of blood vessels in underweight babies develops differently, causing higher blood pressure, with increased risk of stroke and heart disease, as the child matures. There may also be a predisposition to adult diabetes, obesity, schizophrenia and thyroid disorders.

The fact that conditions occurring in pregnancy can have such long-term effects is not surprising when the nature of the developmental process is considered. The millions of cells that make up the adult, twenty or so years after the egg is fertilized at the beginning of pregnancy, are the result of that single cell dividing innumerable times. Because by far the greatest number of these cellular divisions occur in the womb, the greatest exposure to harmful effects occurs during pregnancy.

What is more, the failure of optimal development in early life can lead to long-term changes in the regulation of certain physiological processes, programming the foetus to greater risk of disease. The foetus may respond to undernutrition or other types of suboptimal environment with permanent changes in physiology and metabolism, which may irrevocably change the way the adult body works, resulting in a predisposition for a certain illness.

IN VITRO FERTILISATION

We have already discussed the phenomenon of cloning. It has been said that perhaps the strongest case for cloning a human – apart from the therapeutic cloning we discussed in chapter 6 – is that it could bring hope to childless couples. Many people would sympathise.

But there are methods of initiating a pregnancy (apart from the natural procedure of sexual intercourse) without resorting to cloning. The alternatives are :

- *intrauterine insemination* (artificial insemination), in which sperm is inserted into a woman's uterus and she conceives naturally
- *intrafallopian transfer*, in which unfertilized eggs and specially prepared sperm are introduced into the woman's fallopian tubes. In the most 'natural' version, doctors use the birth mother's egg and her partner's sperm
- *in vitro fertilization* (IVF), in which an egg is fertilized in the laboratory and the resulting embryo transferred into the woman's uterus, or into her fallopian tubes if they are normal. Depending on the origin of the egg and sperm, there is a range of possibilities:
  - IVF using the birth mother's egg and her partner's sperm
  - IVF using the birth mother's egg and donor sperm
  - IVF using a donor egg and sperm from the birth mother's partner
  - IVF using a donor egg and donor sperm
  - surrogacy, where the birth mother carries an embryo conceived using the egg and sperm of a couple who will raise the child after birth
- *intracytoplasmic sperm injection* (ICSI), a new variation on IVF, helpful for men with low sperm counts, in which the sperm is injected directly into the egg in the laboratory. >

> So far, about a quarter of a million babies have been born by IVF worldwide. Even though these procedures are widely used, and are of obvious benefit to many, however, controversy continues about certain aspects of IVF.

One problem is the infiltration of commerce. In some countries, for example, it has been claimed that young female students are paying their way through their tertiary studies by selling their eggs. Desperate couples for whom traditional IVF techniques have failed are being lobbied by commercial brokers to consider eggs and frozen embryos on sale anonymously. Other couples are opting for a 'designer baby' concept, using donor eggs and sperm chosen from a catalogue for their likely looks and /or brains.

These are not the only matters of concern. Young egg donors, for example, face a small but real risk of infertility as a result of overstimulation of the ovaries and related complications; and the recipients of anonymously donated eggs and sperm have children with no traceable genetic linkage.

Some of the procedures are lengthy and emotionally draining, too. IVF, for example, involves a series of hormone injections to stimulate the ovaries, blood tests and ultrasounds to monitor progress, egg extraction, then fertilization and implantation of the embryos into the uterus – which can involve some physical discomfort. It can also be expensive, and it may lead to multiple births. What is more, there is a marked fall-off in the success rate with the age of the recipient – from about 70 per cent if the mother or egg donor is younger than 30, to less than ten per cent if she is over 40.

Debate also continues over the safety of ICSI, with some early studies suggesting a link with an increase in developmental and reproductive problems.

Aberrant development in cancer

To many the link between cancer and aberrant development may not be immediately apparent. Yet the link is strong – so strong that many researchers hope, and expect, that research in this area will ultimately reveal the elusive and long-sought cure for this terrible disease.

We can look at it this way:

- *Development* is a process in which the growth and differentiation of cells are intimately linked and strongly controlled.
- *Cancer* is a process in which the normal growth controls have failed, and the normal processes of differentiation move into reverse. While differentiation is normally accompanied by the appearance of new types of cell, in cancer there is generally a process of *de-differentiation*, or loss of cellular variety.

That is, cell differentiation and tumour growth are both complex biological processes that are widely recognised as having many corresponding, though opposite, characteristics.

We are not going to delve too deeply here into the molecular intricacies of the deficient growth controls in cancer. For now we will just have a brief look at the nature of cancer in general, and the approaches scientists are taking in their search for a remedy.

What is cancer?

Cancer may be any one of more than a hundred diseases characterised by an uncontrolled growth of abnormal cells, which may invade and destroy other tissues. It may be initiated by a complex mixture of environmental, nutritional, behavioural and hereditary factors.

Although cancer is not directly inherited, it is a disease in which genes play a crucial role. In a minority of cases, the affected person has inherited an altered form of a gene that causes a heightened susceptibility to cancer.

Many other cancers are caused by mutations that are not inherited but occur in a person's *somatic* cells (that is, cells other than eggs or sperm) during their lifetime. Such mutations may involve genes called *proto-oncogenes*, which help regulate cell growth. A mutated proto-oncogene may become an *oncogene* – a gene that sparks the continuous unregulated growth that characterises cancer. Oncogenes, for example, contribute to certain cases of leukemia, as well as to ovarian, lung, colon, and other types of cancer.

Then there are *tumour suppressor genes*, which code for proteins that inhibit cell division. Mutations can cause these proteins to be inactivated, so that cells are unable to stop proliferating.

GENES AND BREAST CANCER

Breast cancer is one type of cancer for which a gene increasing susceptibility may be inherited. Two *alleles* (alternative forms of the same gene), each associated with a different gene, have been implicated in some cases of breast cancer. The genes corresponding to these alleles are BRCA1, which is located on chromosome 17, and BRCA2, located on chromosome 13. Women who inherit either of these alleles have a 50 to 85 per cent chance of developing breast cancer during their lifetime, compared to about a 12 per cent chance for others.

Cancer and ageing

Age is also often a factor in cancer. The cells of an older person have typically had more exposure than those of a younger person to the damaging effects of various mutagens, such as tobacco smoke and sunlight. An older person will also have experienced a greater number of cell divisions, offering more opportunities for mutations to occur.

As a result, an older person typically has a greater accumulated store of mutations – a greater *mutational load* – than a younger person. Many mutations have no harmful consequences; others are potentially harmful but are successfully repaired by the cell. Unfortunately, however, some mutations that alter the DNA of proto-oncogenes have the potential to change these genes into devastating oncogenes (see page 139). As a person's mutational load increases, the likelihood of having this harmful type of mutation grows.

Some of the genes involved in human cancers are shown in the box on page 141.

Cellular factors in cancer

Any discussion of cell proliferation and the development of cancer (*carcinogenesis*) on a molecular level must take into account the role of growth factors, since cell division is highly regulated by this family of proteins.

The human body has hundreds of growth factors of different types. Most are *cell type specific*, stimulating division only of certain cells (an example is epidermal growth factor, which promotes the growth of skin); others are more general in their effect. There are also factors outside the cell that work

SOME GENES INVOLVED IN HUMAN CANCERS

	Oncogenes
PDGF	glioma (a brain cancer)
Erb-B2	breast, salivary gland and ovarian cancers
RET	thyroid cancer
Ki-ras	lung and pancreatic cancer
c-myc	breast, stomach and lung cancer
Bcl-1	follicular B cell lymphoma
MDM 2	connective tissue cancer
	Tumour suppressor genes
APC	colon and stomach cancer
DPC4	pancreatic cancer
NF1	peripheral nervous system cancer
P53	a wide range of cancers
BRCA1	breast and ovarian cancer
BRCA2	breast cancer
VHL	renal cell cancer

against the effects of the growth factors, slowing or preventing cell division (these include *transforming growth factor beta* and *tumour necrosis factor*). These are needed to turn off the growth factors where necessary. They act through cell-surface receptors very similar to those for hormones and by similar mechanisms, ultimately leading to an alteration of gene expression.

Growth factors are also known to participate, in some cell types, in differentiation and *apoptosis* (programmed cell death, which removes unwanted cells). Their influence on the processes involved in cell differentiation, proliferation and carcinogenesis is therefore potentially significant.

Earlier we mentioned the genes called proto-oncogenes. When a growth factor message from outside the cell enters it and reaches the nucleus, it activates the proto-oncogenes. When this type of gene mutates, it becomes an oncogene – a gene that instructs the cell to grow and divide repeatedly even if no growth factor is present. Runaway cell division does not necessarily lead to cancer, however. Neighbouring

SOX 18

One gene has been found that promises to stop bowel cancer and other killer tumours in their tracks. This gene, Sox 18, is critical to the formation of the blood vessels needed for the development of brain, colon, bowel and prostate cancers. (Sox 18 may also hold the key to a host of blood-vessel-related diseases, including diabetes and strokes.) Scientists working with this gene are in effect trying to block the blood supply to the tumour cell.

Because the gene inhibits blood vessel growth, it may also slow *metastasis*, the process whereby cancer cells break off from tumours and spread through the body via the circulatory system. Once a cancer has metastasised the chances of recovery drop sharply, so control of the blood supply in this way may be quite critical.

cells may respond by releasing a *growth inhibitor*, a chemical that sends the nucleus of the malfunctioning cell a signal that activates tumour suppressor genes, putting the brakes on cell growth.

Apoptosis is also a topic of much research interest at the moment. If this natural process of cell death could be specifically accelerated in tumour cells, it could offer a major breakthrough in cancer treatment.

Future directions

Scientists, then, have amassed a wealth of information about how cancer works at the molecular level. It is known that cells go through a series of changes before turning cancerous. Once researchers fully understand the molecular detail of those changes, they can design drugs to stop the progress of the cancer.

At present, a broad array of weapons to attack specific properties of cancer cells has been developed. Unlike traditional cancer treatments such as chemotherapy and radiation therapy that often destroy cancer cells and healthy cells alike (and are therefore sometimes referred to as therapeutic blunderbusses), the aim of the new therapies is to act like smart bombs, specifically attacking the defect, causing minimal collateral damage and triggering relatively few side-effects.

Examples of such drugs include therapies designed to prevent growth factors reaching a tumour, and to block signals that would otherwise instruct the cell to grow out of control. Other drugs drive cancerous cells to self-destruct by stimulating apoptosis, while still others block enzymes that cancer cells use to chew openings in normal tissue, and some prevent tumours from building new blood vessels to supply themselves with food and oxygen.

There are many promising leads being followed in present day cancer research.

CHAPTER 10

DNA AND PROTEOMICS

One of the most rapidly advancing areas of research into DNA and the human body is that of *proteomics*. The ultimate aim here is to understand how proteins are synthesized from a DNA template; identify all the important proteins in the human body; determine their three-dimensional structure; and, eventually, understand how they function in the human body in health and disease.

The study of how proteins are constructed and the relationship between their form and their function – *structural genomics* – is likely to be at the centre of medical advances in the next few decades.

Proteins are the major functional elements in the cells that enable the body to carry out its myriad functions. It may be an enzyme that helps you digest your lunch, an antibody that fights off an infection, or one of the receptors in your brain that helps you read and understand this paragraph – they are all proteins.

We have considered the central role of DNA in the production of these proteins, but have not yet described the mechanism of their synthesis. The time has come to revisit protein synthesis – not in fine molecular detail, but enough to understand its general nature and significance.

Proteins and amino acids

Amino acids are organic acids that are the building blocks of proteins.

Proteins are made up from small molecules called *amino acids*, which are linked together in sequences to form *peptides* and proteins. A small protein is shown in figure 10.1.

The amino acid sequence that forms a particular protein determines its individual characteristics, and is in turn determined by the sequence of the nucleotides in the DNA.

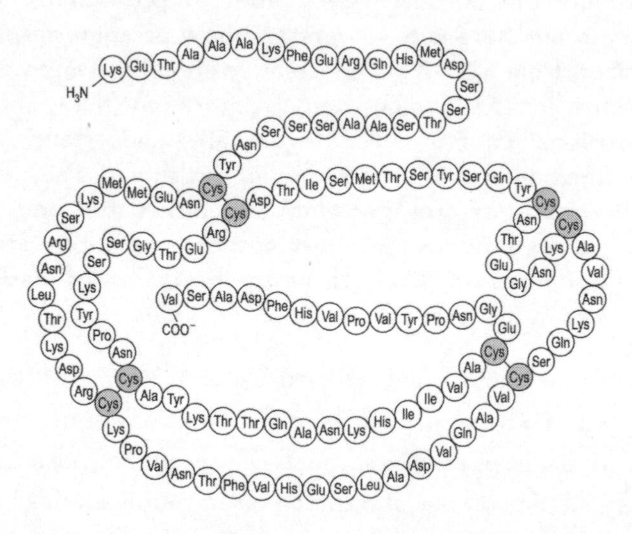

FIGURE 10.1 The amino acid sequence of a small protein. The 124 amino acids of this protein (ribonuclease) are abbreviated for convenience. Larger proteins may contain several times this number of amino acids.

We have seen that DNA contains four different nucleotide bases. These four bases can be thought of as a four-letter alphabet. Like the letters of a linguistic alphabet, which can be arranged into meaningful words, the bases in the nucleotides can be arranged into 'words' that can be understood by the protein-synthesising machinery in the cell. They are all three-letter words; a sequence of three nucleotide bases codes for one amino acid.

Peptides are molecules formed when a small number of amino acids (two or more) are linked together.

It has been known for a long time that the nucleotide bases of DNA carry the basic information required to specify the details of protein synthesis. It was realised early on that the only way just four bases could carry enough information to specify the position and sequence of twenty different amino

MAKING A HUNDRED THOUSAND PROTEINS

There are twenty amino acids that commonly occur in proteins, each consisting of a sequence of three nucleotide bases. To get some idea of the variety this mechanism is capable of producing, consider that a typical protein may contain a hundred amino acids, while a typical molecule of human DNA may contain millions of base pairs. More than enough permutations are possible in this coding mechanism to cover the hundred thousand or so proteins found in the tissues of a human being.

acids was by acting in multiples. Mathematical considerations required these to consist of at least three nucleotides acting in concert; following this lead, biochemists soon afterwards managed to demonstrate that three nucleotides are indeed involved in the specification of a single amino acid.

Proteins are macromolecules formed when a large number of amino acids (usually hundreds) are linked together.

The next question was which nucleotide groups, or *codons,* coded for which amino acid; and this was soon established too. The detail of this code is shown in the box on page 149.

Transport of nucleic acids in the cell

Another problem for nature to deal with in managing protein synthesis in mammals concerns the location of the various cellular components and processes. DNA is located mainly in the nucleus, and restricted to that location by the surrounding nuclear membrane, whereas most cellular proteins, and protein synthesis, occur outside the nucleus in the cytoplasm. A mechanism to cope with this – a means of transporting the necessary information for protein synthesis from the nucleus to the cytoplasm – is required.

In the nucleus, one of the strands of DNA becomes a template for the formation of another form of nucleic acid – messenger RNA (mRNA). We mentioned this briefly in chapter 1. This process is similar to that of DNA duplication observed during cell division, except that it uses a different enzyme (called *RNA polymerase*) and the base *uracil* instead of thymine.

After the genetic information has been transcribed, the mRNA, which is a very much smaller molecule than DNA,

THE GENETIC CODE

The 64 triplet codons (sequences of nucleotide bases) are listed below, with the abbreviated term for the amino acid that is encoded by each triplet. The four nucleotide bases are represented as usual by the letters U, C, A and G. Most amino acids have more than one possible codon. The codon that signals the termination of a protein sequence is indicated as 'term'.

UUU	Phe	UCU	Ser	UAU	Tyr	UGU	Cys
UUC	Phe	UCC	Ser	UAC	Tyr	UGC	Cys
UUA	Leu	UCA	Ser	UAA	term	UGA	term
UUG	Leu	UCG	Ser	UAG	term	UGG	Trp
CUU	Leu	CCU	Pro	CAU	His	CGU	Arg
CUC	Leu	CCC	Pro	CAC	His	CGC	Arg
CUA	Leu	CCA	Pro	CAA	Gin	CGA	Arg
CUG	Leu	CCG	Pro	CAG	Gin	CGG	Arg
AUU	Ile	ACU	Thr	AAU	Asn	AGU	Ser
AUC	Ile	ACC	Thr	AAC	Asn	AGC	Ser
AUA	Ile	ACA	Thr	AAA	Lys	AGA	Arg
AUG	Met	ACG	Thr	AAG	Lys	AGG	Arg
GUU	Val	GCU	Ala	GAU	Asp	GGU	Gly
GUC	Val	GCC	Ala	GAC	Asp	GGC	Gly
GUA	Val	GCA	Ala	GAA	Glu	GGA	Gly
GUG	Val	GCG	Ala	GAG	Glu	GGG	Gly

moves through the pores of the nuclear membrane to the cytoplasm, where it attaches to small bodies called *ribosomes.* The shape of a ribosome allows it to be threaded onto the mRNA like a bead on a necklace, and it passes along the mRNA reading the code for the particular amino acids required for protein synthesis. These amino acids are then joined together in a long chain (a *polypeptide*) whose sequence is specified by the mRNA. When it is the right length, an amino acid called a *stop codon* terminates synthesis and releases the protein to take up its position in the cell.

Thus the DNA manages to accomplish two essential but very different functions – protein synthesis and its own replication – in a simple yet elegant fashion.

The human proteome

Even before the human genome project was completed, it was considered passé by many scientists. They had begun thinking forward to the even greater challenge of explaining the composition and function of the human body in health and disease. This means, most critically, identifying and characterising the human *proteome* – the entire complement of proteins encoded by the genome.

The task is monumental. The process of translating genes into proteins in humans is now known to be more intricate than was once thought. This is because in humans, some genes can be read in more than one way, and the protein products are subject to modification. So although the number of human genes is comparable to that in many simpler organisms (around thirty thousand), the number of human proteins is

THE GENOMES OF MODEL ORGANISMS

With the human genome project near or at completion, many laboratories are giving their attention to determining the genomes of much simpler organisms, such as yeast, worms, fruit flies and mice. While at first there might not seem to be much connection between these efforts and the human genome project, in fact they are proving very useful in the study of a variety of human diseases.

The genes of these *model organisms* are attracting attention because in many cases the proteins they encode closely resemble those of humans – and of course such organisms are much more practical than humans to work with under controlled laboratory conditions.

One genome that has been sequenced is that of the fruit fly *Drosophila melanogaster*. Scientists have found that half the fly proteins show similarities in molecular shape to human proteins. One fly gene producing a protein similar to human's is p53, a tumour-suppressor gene that, when mutated, allows cells to become cancerous. The p53 gene is part of a molecular pathway that causes cells that have suffered irreparable genetic damage to commit suicide. Similarities such as this allow scientists to do very sophisticated genetic manipulation in studying such a pathway – manipulations they cannot do in larger animals because of the expense, and the length of time between generations.

Of course, the results must ultimately be tested in mammals, and that generally means laboratory mice and rats. It is interesting, then, that more than 90 per cent of mouse proteins identified so far show similarities to the corresponding human proteins.

considerably larger (probably in the order of hundreds of thousands, although nobody knows exactly).

Proteins, too, are far more complex in their structure and function than DNA. They are composed from twenty amino acids, not four bases like DNA, and they are far more variegated and complex than DNA. They have to be. Every chemical reaction that is essential to life depends in one way or another on their services. Proteins provide the structural elements of the cell, and the means of holding cells together to form tissues. They are the enzymes that drive the thousands of reactions that allow our cells to function properly, many of the hormones that direct metabolism throughout our bodies, and the antibodies that guard against infection.

What a protein does in terms of its cellular function is largely determined by its shape – not its outward appearance, but its detailed three-dimensional structure. The external surface of a protein is pitted with clefts into which smaller molecules fit like a key in a lock, and these precise interactions determine what metabolic reactions occur, and to what extent.

To fully understand how a protein works, then, scientists have to know both the exact sequence of its amino acid components and their precise three-dimensional arrangement.

When you consider how may different proteins there are, determining the human proteome emerges as a far larger task than the 'gigantic' human genome project. In a sense, that project was merely the ceremonial start to an organised world-wide program to attack disease by cataloguing the essential molecular elements of human biology.

Despite the magnitude of the task, determining the details of the human proteome is a target that scientists – and pharmaceutical companies – are pursuing with great vigour and purpose.

DRUGS BY DESIGN

Knowing the exact structural form of each protein in the human proteome should in theory enable the design of drugs in the form of chemicals that will fit the slots on any particular protein and, for example, either activate it or prevent it from interacting.

This is exactly what was done with the anti-influenza drug Relenza. Having determined the precise three-dimensional structure of an enzyme that was essential for the multiplication of the flu virus, scientists were able to design an inhibitor that fitted into a slot in the enzyme protein, nullifying its activity and preventing the virus's normal proliferation.

The challenge at the moment is to apply this type of approach to the hundreds of human proteins involved in disease states. At present, however, the structure of only a few per cent of all human proteins has been determined, so in the meantime scientists often use a more roundabout method: they look for a gene in a model organism that is similar to a given human gene, translate the gene into an amino acid sequence using a computer program, look for similar sequences in databases of model organism proteins, model the human protein based on the model organism protein's known structure, then look for a drug that binds to the modelled protein.

In one sense, the process of defining the human proteome is not new. Scientists worldwide have been involved in determining protein structure over several decades; but progress has been slow by comparison with that in the human genome project. With the success of the human genome project behind us, however, the pressure is on to greatly increase the rate of discovery; and the key, just as with the human genome project, appears to be automation. Here we will briefly consider the main elements involved in protein analysis.

Protein analysis

Up to now, protein analysis has involved a number of tedious, time-consuming steps:

- separation of the proteins by, for example, *chromatography* or *electrophoresis* (processes that separate proteins on the basis of their size, shape or electronic charge), then
- chemical analysis of a protein's components (identifying the amino acids and other components), then
- sequencing of the protein (determining the order in which the amino acids are linked), then
- working out its three-dimensional structure.

This sequence of steps is easy enough to describe in words, but in practice it may turn out to be a truly monumental task; the analysis of a single protein has, in the past, often been beyond the capacity of one scientific lifetime. Analysis has been more like a craft than an industrial process. So scientific organisations are moving towards computerisation to speed up the flow of information a thousandfold or more.

NEW METHODS IN PROTEIN ANALYSIS

A number of different types of machine are on the drawing board. One type would use electrophoresis to separate proteins, automatically extract individual proteins from the gels, split the proteins into pieces with enzymes, feed the pieces into a *laser mass spectrometer* and transfer the information so obtained to a computer for analysis.

Another method involves the so-called guilt-by-association technique; learning about the function of a protein by assessing whether it interacts with another protein whose role in the cell is known. In this way, proteins may be sorted into functional categories according to their role in, for example, energy generation, DNA repair or the ageing process.

There are also new, faster methods for determining the relationships between individual proteins and disease. One approach is to use a bank of genes (a *gene chip*) that is big enough to synthesise most of the proteins in the human body and test the extent of production of these proteins in living cells. This allows the performance of *parallel tests*, which compare the expression of the proteins in normal and diseased cells. This, in turn, enables scientists to determine the roles of the proteins in, for example, cell division (important in cancer) or the way cells move about in the body (important in wound closure).

Yet another approach is to compare the proteins in diseased and healthy tissues, and in diseased tissues before and after treatment, and try to work out how >

> diseases affect proteins at a molecular level. Gene chips can be used for this purpose; so can special molecular *reagents* that allow the separation of particular classes of protein so that they can be examined in more detail. In this way the amount of a protein that has been affected by the disease can be ascertained, and the protein in question can be extracted in bulk and analysed further to detect any modifications caused by the disease.

It is hoped that the range of new techniques will advance research in this area many times more rapidly than in the past.

Structural genomics

The area of proteomics concerned with identifying the molecular shape of particular proteins is called *structural genomics.* The aim in the next decade is to describe the shapes of about ten thousand of them. As we have seen, that is only a fraction of all the proteins in the human body, but scientists believe that this number will cover most of the protein structures that have significant roles in disease.

Why not analyse them all? For one thing, there are just too many. But in any case, it probably isn't necessary. Scientists are reasonably sure that most of the hundreds of thousands of proteins in the human body are variations on a handful of central themes. Proteins with a similar function in different life forms often share structural characteristics that are reflected in the genes that encode them. Using this knowledge, scientists hope to construct a catalogue of shapes that when mixed and matched will spell the shape of any gene's product.

Of course, identifying the general shapes of classes of proteins is one thing; knowing exactly how they form into specific proteins after much folding and linking is another. Scientists have traditionally discovered these details by extensive purification of the protein, then measuring how the crystallised protein scatters X-rays. Such experiments can take years to complete; again, automation will speed the process. Specialised machines such as the *synchroton* have already been developed to facilitate these analyses.

You are going to be hearing a lot more about such machines, and about proteomics, in the next few years.

CHAPTER 11

DNA AND THE FUTURE

There is no doubt that humanity is in the midst of a major technological transition – one that will have effects comparable to those of the industrial and computer revolutions, and that will lead us into a biotechnological century. We are entering an era of genomic medicine, for example. Emphasis is shifting from the genetic basis of disease to a more detailed understanding of cellular function in health and disease, with the role of proteins being at the forefront of these advances. Huge databases are being constructed, opening up the new field of bioinformatics in which many aspects of diagnosis and therapeutic treatment may be decided on the basis of computerised information. Our understanding of the functioning of the human body is growing at an ever increasing rate, and these momentous scientific advances are raising new questions in the field of ethics.

Many of these discoveries are so recent that they were not even hinted at when most people who are now adults were at school, college or university. They are of such wide-ranging consequence, however, that it is essential we know something about them. We need to be on more

familiar terms with the biology of DNA, so we can make responsible judgments on the future interactions between biotechnology and society.

DNA itself is not the end of the biotechnology story. The deterministic view that the genetic blueprint makes humans a simple matter of cause and effect – that the biological fate of an individual is totally determined by their genes – is now well and truly discredited in serious scientific circles. The function of the human cell (and the human body) is far more complex than that. While discussing the role of DNA and genes might be fashionable just now, the emphasis of biotechnology is turning away from genes as such to the way in which those genes perform within cells and tissues; and there is a subtle change in the thrust of biological research, with the central concern moving from molecular biology to cell and tissue biology.

Perhaps this change is best exemplified by the status of the human genome project. Now that many of the genes that make up the human genome have been decoded, the emphasis is turning from questions about the chemical nature of our genes to questions about other elements of the biological processes that define our identity. The DNA sequence of the human genome tells only a small part of the story about what a specific cell is doing. Questions are posed not only about the nature and function of the immediate gene products – proteins – but also about the way in which proteins interact to influence cellular metabolism.

This change of emphasis does not deny the central role of DNA, but rather extends it. It highlights the fact that the functioning of the human body is complex, and that scientific understanding of that functioning still has a long way to go.

The rise of genomic medicine

Recent advances in our knowledge of DNA have the potential to lead to great improvements in the human condition. Prominent among these is the promise of *genomic medicine*. It is anticipated that knowing the details of the human genome will enable scientists to find the genetic triggers for hundreds of human diseases; to devise extremely sensitive diagnostic tests for these conditions; and to develop drugs that are specific, and effective, in their treatment. This could change the present conception of the nature of disease, drastically improve diagnostic procedures, and lead to much more rapid and effective treatment.

Having the DNA sequences available in their entirety gives scientists a long head start into the era of genomic medicine, enabling the use of tools such as gene chips that allow them to identify much more rapidly than in the past which of the thousands of genes in a tissue are active at a given time.

Understanding multiple gene activity

The activities of single genes make a lot more scientific sense when considered in the context of the action of other genes. As one scientist has said:

> Right now it's like watching a movie on a few pixels at a time and trying to figure out the whole story. Having the complete genome sequence gives you the whole image on your screen, and should result in a substantial increase in the rate at which discoveries are made.

GENOMIC MEDICINE AND CANCER DIAGNOSIS

The usual procedure at present for diagnosing cancer is for a pathologist to look at a tissue sample under a microscope to identify the presence of a tumour. This procedure is far from infallible. The sample may have actually originated from a quite different tissue source, or contain more than one type of tumour. Such situations are much more amenable to analysis and treatment through a knowledge of gene sequences, which can give a more reliable indication of the source and type of primary tumour. Genomic medicine seems set to play an increasingly important role in this and many other disease states.

Yes, the age of genomic medicine has arrived, and the sequencing of the human genome represents its formal beginning. A few decades from now, our understanding of the human organism and its various ills is likely to be transformed beyond present day recognition.

The big picture

The pinpointing of the genetic basis for disease is only a minor part of what genomics is all about. The central issue today is no longer genes but what is done with them. It is necessary to understand the other elements of the biological process and link all the information together.

Following on from DNA are its biological products –

mRNAs and proteins. If DNA is the set of master blueprints a cell uses to construct proteins, mRNA is the copy of part of the blueprint that the builder takes to the construction site every day. DNA remains in the nucleus of the cell; the mRNAs transcribed from active genes leave the nucleus to give the orders for making proteins in the further reaches of the cells.

As we know, although every cell in the body contains all the DNA code for making and maintaining a human being, many of the genes are never turned on once embryonic development is complete. Various other genes are turned on or off at different times – or not at all – according to the tissue they are in and their role in the body. A *pancreatic beta cell*, for example, is generally full of the mRNA needed to make insulin, whereas a nerve cell in the brain has little or none. A full explanation of this process of differential gene action in mammalian tissues remains a major aim of research into human biology.

Scientists used to think that one gene equates with one protein, but the reality is much more complicated. We now know that one gene can be read out in portions that are spliced and diced to make a variety of mRNAs, and that subsequent processing of the newly made proteins which those transcripts encode can alter their function.

The DNA sequence of the human genome therefore tells only a small fraction of the story about what a specific cell is doing. Researchers must also pay attention to such matters as the *transcriptosome* – the body of mRNAs being produced by a cell at any given time – and the proteome, all the protein being made according to the instructions in those mRNAs.

'Junk' DNA

We explained earlier that most of the DNA in humans and other mammals – more than 90 per cent – is in the form of 'junk' or non-coding DNA. The significance of this type of DNA was not obvious to early workers. Most of the scientific attention was directed towards the sequences of coding DNA, and non-coding DNA was often explained away as an evolutionary vestige.

Our understanding of this form of DNA is increasing, however, largely as a result of the recent availability of detailed sequence data for the genome of humans and other mammalian species.

It has become clear, for example, that non-coding DNA contributes much more than coding DNA in defining differences between species. Humans and chimpanzees have 98.5 per cent of their coding DNA in common, for example. They have far less non-coding DNA in common – a fact that may give comfort to those people who dislike the idea of so much similarity between humans and apes.

The role of non-coding DNA

Some of the most exciting developments with this type of DNA concern the possibility that it may have a central role in differentiation and development.

Recent sequencing of mammalian genomes has shown that along with the interspecies differences, there are also major areas of identity. The fact that evolution has not tampered with these sequences suggests very strongly that they are absolutely essential in mammalian development.

Their role must be very different from that of the coding DNA, however, since the non-coding sequences are by definition not directly involved in coding for proteins. Instead they make a type of RNA that is different from mRNA. Some scientists now believe that these sequences of non-coding DNA act as a master regulator, switching on and off the differential protein synthesis that is an essential factor in the growth and development of all mammals.

Investigations such as these hold great promise for furthering our understanding of the role of DNA in health and disease, and will be followed with great interest.

The DNA industry

Another important, though different, aspect of the massive increase in available information on human DNA is the new industry that is emerging to capitalise on this data.

We have mentioned the use of DNA information in diagnosis and drug development. This has so far attracted billions of dollars in venture capital and other financing. The next cab off the rank involves proteomics. Most present day pharmaceutical drugs are derived from natural products, and this has been the major emphasis of research into new products; but that is about to change radically. Expensive trial and error is being replaced by the production of drugs by design – that is, using the detailed information about genes and gene products to produce a desired interaction between the designer drug and specific functional proteins in human tissues.

The principal steps in the sequence are shown in figure 11.1.

Using gene chips, scientists can now identify the activity

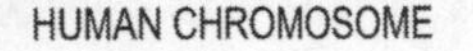

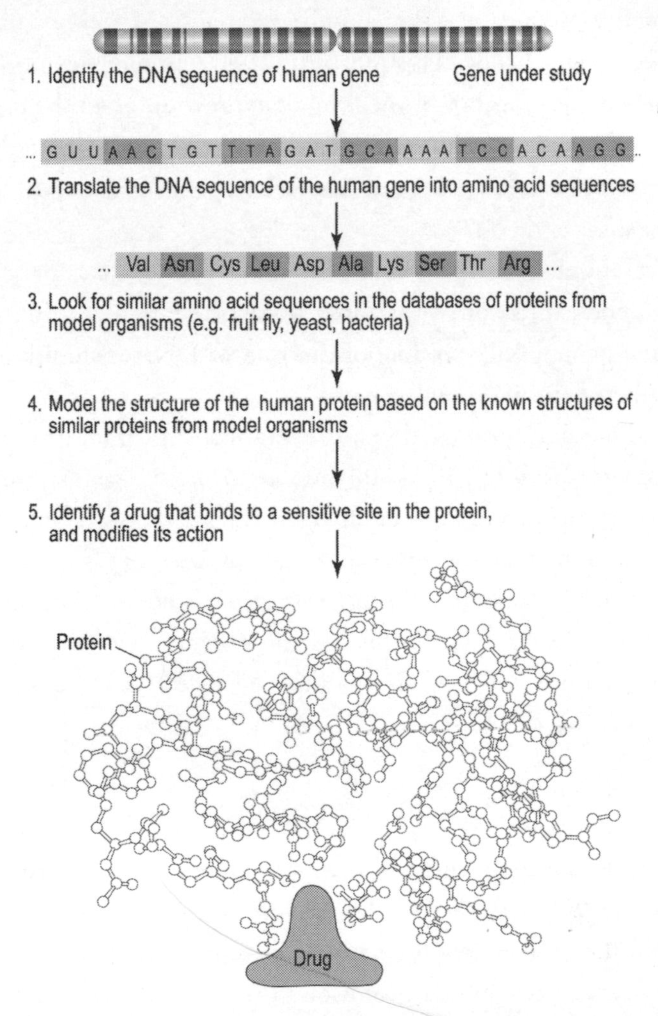

FIGURE 11.1 How DNA sequences may be utilised to identify promising drug treatments for human diseases. The sequence information from humans is compared with the more complete data available for model organisms, to deduce a probable structure for the human gene product. This facilitates the search for appropriate drugs.

of tens of thousands of mRNAs involved in the synthesis of specific proteins at a rate hundreds of times faster than was previously possible. This powerful new technology can give a convenient snapshot of the level of expression of all the genes in a cell, and is enabling rapid advances in our understanding of the molecular processes that occur in differentiation and disease.

Gene chips can also be used to monitor the effects of drugs on gene expression, and in particular their influence on the proteins in a cell's production line – which, after all, are generally the ultimate drug targets.

It has been estimated that the pharmaceutical industry will evaluate some ten thousand human proteins against which new therapeutics might be directed within the next decade – that's more than twenty times the number of drug targets reached by all pharmaceutical companies since the beginning of the industry. Clearly this has the potential to increase the drug output (and profit margin) of many pharmaceutical companies.

Bioinformatics

Another development in this area is *bioinformatics*. There are now gigantic databases available on nucleic acid sequences from the human genome project, as well as a large number of sequences from model organisms. Data are also being generated about when and where in the body these genes are being turned on, the shapes of the proteins they encode, how the proteins interact with one another, and the role those interactions play in disease.

DNA AND MABS

Another example of the clinical use of proteins and the new technologies is the use of *monoclonal antibodies* (MABs) for therapeutic and diagnostic purposes.

Antibodies are proteins that constitute one of the body's primary defences against the myriad alien substances that we encounter during our lifetime – microbes, parasites, toxins and so on. Recently it has become possible to copy the DNA (and RNA) sequences of the human immune system that are concerned with the synthesis of specific target *antigens* (substances that trigger the production of antibodies), and by inserting these sequences into an appropriate host organism to synthesise MABs in industrial quantities. They can then be used to seek out and destroy particular viruses or bacteria, or to correct metabolic disorders.

MABs can also be used for diagnostic purposes such as providing an early warning of the presence of cancer, identifying metabolic errors, and confirming the specific nature of microbial infections.

There are mountains of data available. The new discipline of bioinformatics – a marriage between computer science and biology – seeks to make sense of it all. In so doing, it is destined to change the face of biomedicine.

First of all, however, the enormous quantity of the available data needs to be emphasised – at present, there are some hundreds of thousands of CDs full of scientific data waiting to be correlated – and the question arises as to where this work will lead.

We may well be at the beginning of a revolution in which scientists will be able to use bioinformatics to build computer models of the astronomical numbers of biochemical reactions that add up to a human life – not only the functions of individual genes but also the dizzyingly complex network of enzyme reactions, receptor interactions and protein binding patterns that result. They will be able to model not only the building blocks of human life, in short, but the entire working machine – a virtual cell within a computer. This, in theory, will allow researchers to test potential drugs for safety and effectiveness in vitro, and it represents a fantastic drug-discovery engine for the future.

Beyond that, the biggest bioinformatics project in front of us is unscrambling the regulatory network that controls cell development from the fertilized egg to the adult – a key development, as we have seen, in understanding and extending the applications of biotechnology.

Into the future

The influences of biotechnology are seen by many as affecting the well-being of humanity far beyond genomic medicine. We have already discussed the ability of new technologies to increase the pace of biological discovery by some orders of magnitude, and it has been predicted that bioinformatics will change our world more dramatically than any other technological revolution in history, leading to economies based more on genetic commerce than oil or coal or any other present day resources.

Just as the computer age has affected all economies, so the influence of biotechnology (including the marriage of genomics and informatics) will have worldwide effects on economies and the human condition. Biotechnology is poised for a massive global expansion, and as a key driver of economic growth it will affect every aspect of our lives – not only our medical treatment and our environment, but the food we eat, the way we raise our children, our work patterns, and which societies thrive and which founder.

Ethical issues

It is not too much to say that human life as we know it is changing beyond recognition with the scientific advances of gene technology. Many of these raise significant ethical issues, and some deserve to be at the top of the public agenda.

Privacy

One issue raised by our ability to obtain information on DNA sequences is the question of privacy. There will be a real temptation to pry into each other's genomes. Say your employer has a blood sample taken for a physical examination, but decides to check risk factors for manic depression.

Laws are needed guaranteeing the absolute privacy of a person's genomic information in such situations. You may choose to waive your privacy – for example, if you want to get special insurance for cancer risk. But the choice should be yours.

Genetic determinism

Then there is the question of genetic determinism. Without a background in the subject, people tend to oversimplify genetics, saying that someone has a 'gene for cancer' or a 'gene for diabetes'. But the fact is that genes determine only so much. Identical twins have identical genomes, yet one may develop juvenile diabetes while the other does not, for example.

People tend to invest genes with a power over the future that is simply not justified by science. Understanding the role of genes should help us to pinpoint environmental factors and teach us to avoid the consequences of our genetic predispositions.

The right to genetic information

There are questions about the right to patent genetic information. At present, it seems that a patent may be obtained with only a fragmentary description of a gene's structure, and with little idea at all of its function. This would seem to limit and discourage much future research into how to use the genes to cure disease, or even more modest applications – eliminating the side effects of drugs, for example, by screening them against a complete set of human proteins.

Germ line modification

There is the question of *germ line modification* – whether it is right to modify the genetic code in such a way that people pass on particular modifications to their children. There is a deep moral issue involved in this, in the view of many. Some people say that once human beings are seen as a product of manufacture a line is crossed, and return may be impossible. Many would like to see a ban on modification of the human germ line, for the present. Society could choose to repeal the ban if it ever became technically adept enough – and morally wise enough – to use this power properly, but until then there should be a strong presumption against tampering.

Ethical issues and public opinion

Some of these ethical issues have already been tested by public opinion polls in the US and other countries. Public opinion seems to be clearly in favour of individuals having access to their own genetic code and its medical implications, but against governments or health insurers having this information.

Generally, an appreciable majority are prepared to compromise on ethical concerns about advances in biotechnology as long as they lead to benefits in health, medical care and the environment. In this view, the potential for good outweighs the concerns of people who are seen as alarmists. The biggest reservation among those polled was about genetically modified foods, but even there nearly 70 per cent felt that the food on their supermarket shelves was adequately tested and safe, and if solid science deemed a biotechnology product good for your health and safe to use, then science should trump any ethical considerations. Most people also reject the notion that such technology interferes with the natural order of things.

Looking ahead

Advances in biotechnology over the past few years have been both spectacular and multifaceted, and the coming decades promise even more scientific pyrotechnics.

While the human genome project has held the headlines for some years, it is now being overtaken by investigations into the proteome; but even this project should only be viewed as a stepping-stone towards the broader objective of a full understanding of human cell biology at the molecular level.

If a unifying conception of this combination of projects is required, it may well be found in the ideas of human development and differentiation. Not only do these processes define the essential biological elements that establish our identity, they also encompass most of the areas in which biomedical science is progressing in its endeavour to define the workings of the human body – the differential synthesis of proteins, cellular control mechanisms, protein interactions, and so on.

These matters are complex, but science is making giant strides towards their ultimate resolution – in some cases, the rate of progress has increased by several orders of magnitude in recent years.

The impact of advances in DNA technology is of tremendous importance to our society and our future. Only an informed public is going to be in a position to make appropriate decisions on how to use them. It is clear that the world needs not only politicians, but also a populace, that is familiar with the essential elements of these advances – people who have a broad grasp of the basics of DNA biology and biotechnology – in order to make sound decisions on these important emerging issues.

GLOSSARY

The language of modern biology and biotechnology

Adrenal gland a gland located over the kidneys that releases the hormone epinephrine

Adsorption a process by which a substance forms a film or layer on a solid surface

Allele one of two or more alternative forms of the same gene

Amino acids organic acids that are the building blocks of proteins

Amniocentesis a prenatal screening process that involves analysing a sample of the amniotic fluid

Amniotic fluid the fluid that surrounds the foetus in the uterus

Antibiotics organic compounds that are formed in micro-organisms and plants, and can be used to kill some disease-causing bacteria

Antibody a protein synthesised by the immune system of vertebrates as part of the body's defences against foreign substances

Antigen a molecule that causes a particular antibody to be synthesised

Antioxidant a substance that inhibits the destructive oxidation action of oxygen free radicals

Apoptosis a form of *programmed cell death*, a regulated process by which unwanted cells (whether damaged or unwanted for some other reason) are destroyed

Asexual reproduction any form of reproduction that involves only one parent. Examples are reproduction by cell division in a single-celled animal, and growing a plant from a cutting

Base one of the chemical constituents of nucleic acids. The bases in DNA are adenosine, thymine, guanine and cytosine. In RNA, uracil replaces thymine

Base pair two nucleotides that are linked through their complementary bases. Adenine is paired with thymine, while cytosine is paired with guanine

Biodegradable able to be broken down by the action of living organisms

Biodegradable polymer large synthetic molecules able to be broken down by the body

Biomolecule An organic compound normally present as an essential component of living organisms

Blastocyst an early stage of the embryo before differentiation takes place

Cancer one of a number of diseases characterised by an uncontrolled growth of abnormal cells, which may invade and destroy other tissues

Carcinogen a cancer-causing agent

Carcinogenesis the development of cancer

Carrier [of a genetic disease] a person with a single copy of a faulty gene which would, if the person had two such genes, cause them to develop the disease. Carriers typically show few symptoms of illness

Catalyst any substance that initiates or accelerates a chemical reaction

Chimera an organism carrying genetically distinct tissues derived from two or more different species

Chromatography a technique for separating a mixtures into its constituent parts on the basis of their affinity with a particular substance

Chromosomal disorder a disorder characterised by an abnormality in the number or structure of the chromosomes

Chromosome a single large DNA molecule and its associated proteins, containing many genes

Clone an organism whose DNA is identical to that of another organism

Cloning the production of large numbers of identical DNA molecules or cells from a single ancestral DNA molecule or cell

Codon a sequence of three adjacent nucleotides in a nucleic acid that codes for a specific amino acid

Chorionic villus sampling a prenatal screening process that involves analysing a sample of the *chorionic villus* (part of the placenta)

Cytoplasm the portion of a cell's contents that is within the plasma membrane but outside the nucleus

Cytoskeleton the network of filaments that provides structure and organisation to the cytoplasm

Cytosol the fluid part of the cytoplasm, consisting of a concentrated solution of proteins, RNA and other substances

Deoxyribonucleic acid the full chemical name for DNA

Dermis the inner layer of the skin

Developmental pathway [of an organ] the processes by which the organ grows and develops

Differential gene function the rate at which individual genes are expressed at different stages of development

Differentiation specialisation of cell structure and function during embryonic growth and development

DNA expression the way in which the genetic information carried by the DNA is manifested in the structures and functions of a cell

Dominant gene a gene which exerts its full effect when only one copy is present

Dopamine a chemical that transmits signals in the brain

Double helix the coiled conformation of two complementary nucleotide chains

Ectoderm the outermost of the three *primary germ layers*

Electrophoresis a group of techniques used to separate molecules according to physical characteristics such as electrical charge

Embryo an organism in the earliest stages of development (in humans, the embryonic stage occupies the first seven weeks of a pregnancy)

Embryo splitting a cloning technique that involves dividing an embryo whose cells have not yet begun to differentiate

Endoderm the innermost of the three *primary germ layers*

Endoplasmic reticulum an extensive system of double membranes in the cytoplasm of eucaryotic cells

Enucleation the process of removing the nucleus from a cell

Enzyme a protein that initiates or accelerates (that is, *catalyses*) a particular chemical reaction

Epidermis the outer layer of the skin

Epigenetic influenced by environmental factors that affect gene expression during development

Epinephrine a hormone released by the adrenal glands, which helps the body respond to stress

Eucaryote A single-celled or multicellular organism whose cells have a membrane-bounded nucleus, chromosomes and organelles. This includes all plants and animals, but not bacteria or viruses

Expression *see* gene expression; DNA expression

Fatty acids a class of chemicals present in animal and vegetable fats and oils

Foetal alcohol syndrome a condition caused by a heavy maternal alcohol intake in pregnancy, and characterised by a range of intellectual, emotional and behavioural disorders

Foetus an organism in the post-embryonic stages of development, in humans from the third month of pregnancy until birth

Free radical *see* oxygen free radical

Gastrula the stage in embryonic development following the blastocyst, when the cells are beginning to differentiate

Gene a segment of DNA that codes for a particular protein, RNA molecule or control function

Gene chip a glass or silicon chip that contains fragments of DNA, used to test a sample for the presence or activity of particular genes

Gene expression the process by which the information carried by a gene is converted into the gene's product

Gene therapy treatment of genetic disorders by the replacement of a faulty gene with a normal gene sequence

Genetic determinism the belief that all aspects of a person's physiology and behaviour are controlled by their genetic make-up

Genetic engineering the process of manipulating genes to produce a specific result

Genetically modified organism an organism whose genetic material has been deliberately altered to produce particular characteristics, sometimes by the introduction of genes from another species

Genome all the genetic information encoded in a cell; the total genetic information for a particular individual or species

Genomic medicine medical use of the knowledge of gene sequences for the prevention, diagnosis and treatment of disease

Germ line cells reproductive cells including eggs and sperm

Germ line therapy gene therapy that affects the germ line cells, so that the changes are passed on to subsequent generations

Gland an organ that releases substances such as hormones to be used elsewhere in the body

Glucagon a hormone important in glucose metabolism

Glucose a form of sugar that is the main source of energy in most animal cells

Glycogen the main form in which glucose is stored in animal tissue

Glycolysis the breakdown of glucose to produce energy

Green revolution a process of technological development in agriculture that began in the 1940s with the object of increasing efficiency and productivity, thus helping developing countries meet the needs of growing populations

Growth factor a protein that directs cell growth and differentiation

Growth inhibitor a chemical that initiates the activity of tumour suppressor genes

Haemoglobin the main protein found in red blood cells

Hepatocyte the major cell type of liver tissue

Homeostasis a state of functional and chemical equilibrium in living organisms

Hormone a chemical synthesised by a gland and carried by the bloodstream to a target tissue where it acts to regulate tissue function

Hormone receptor a protein that binds to a specific hormone and initiates its action

Hydrolytic enzymes enzymes with the ability to split DNA (and some other) molecules

Hybridisation the binding of strands of DNA through their complementary bases

Hydrolysis a chemical process in which a molecule is divided into parts by taking up the elements of water

Hypothalamus a region of the brain concerned with the regulation of various metabolic processes

Immune response the capacity of a vertebrate animal to generate antibodies against a foreign protein

Immune system a network of interacting systems and processes involved in defending the body against foreign substances

Immunoisolation therapy a form of treatment whereby cells are implanted into a recipient inside a plastic capsule that shields the cells from attack by the recipient's immune system

Insulin a hormone released by the pancreas, important in glucose metabolism

Intracytoplasmic sperm injection (ICSI) a fertilisation technique in which sperm is injected directly into an egg in the laboratory

Intrafallopian insemination a form of assisted reproduction in which sperm and eggs are introduced into the fallopian tubes

Intrauterine insemination artificial insemination, in which sperm is introduced into the uterus

In vitro fertilisation (IVF) a technique involving the fertilisation of an egg outside the body, with the resulting embryo subsequently being implanted into the uterus

IVF *see* in vitro fertilisation

'Junk' DNA *see* non-coding DNA

Ketone bodies chemicals produced in the liver, whose presence in the urine is one symptom of diabetes

Laser mass spectrometer a device for separating different types of molecules on the basis of their differing mass

Lipid a group of water-insoluble biomolecules, including the fats

Liposome a fatty envelope used to introduce genetic material (or other substances) into a cell

Lysosome a membrane-bounded organelle in eucaryotic cells containing a number of hydrolytic enzymes

Mesoderm the middle layer of the three primary germ layers

Messenger RNA (mRNA) a class of RNA molecules, each of which is complementary to one strand of DNA, and carries the genetic message of that strand from the nucleus to the cytoplasm

Metastasis the process by which cancer cells are transported from a tumour to other sites in the body

Mitochondria organelles in eucaryotes that contain the enzymes for a number of oxidative reactions, as well as some DNA

Mitosis the process of cell division and the replication of chromosomes in eucaryotic cells

Model organism a relatively simple organism that is studied in order to gain insight into the workings of more complex organisms

Monoclonal antibody an antibody produced in a laboratory to bind to a particular receptor site, used for diagnosis or treatment of some diseases

Morphogenesis the structural development of an organism

Morphology physical form or structure

Multifactorial disorder a disorder caused by the influence of several genes as well as environmental factors

Mutagen factors in the environment capable of causing mutations in DNA

Mutation a permanent change in the structure of a gene

Mutational load the accumulated store of mutations carried by an organism

Myosin a protein involved in muscle contraction and in the transporting of organelles in the cytoplasm

Neuron a cell type that is specialised for the transmission of nerve impulses

Non-coding DNA sections of the DNA that do not contain genes (about 97 per cent in humans). It used to be called 'junk' DNA; however, it is becoming increasingly apparent that it has important functions

Nucleic acid DNA or RNA

Nucleotide an organic molecule consisting of a base, a sugar and a phosphate. DNA is composed of long strands of nucleotides, generally in pairs

Nuclear transfer a technique used in cloning that involves introducing the nucleus of a donor cell into an egg cell whose nucleus has been removed

Nucleus an organelle in the cells of eucaryotes that contain the chromosomes

Oncogene a gene that causes cancer

Organelle a membrane-bounded structure in a eucaryotic cell that contains the components required for specialised cell functions

Oxidation a type of chemical reaction in which oxygen, or oxygen free radicals, combine with other substances, which can cause damage to the body's cells and to DNA

Oxygen free radicals highly reactive chemicals that are a by-product of metabolism

Pancreas the gland that releases digestive fluids, and the hormone insulin

Peptide two or more amino acids linked together

Peroxisomes organelles in eucaryotic cells that are important in the removal of oxygen free radicals and other metabolic functions

Pituitary gland a gland situated at the base of the brain that releases hormones relating to a wide range of bodily functions

Plasma membrane the exterior membrane enclosing the cytoplasm of a cell

Pluripotent having the potential to become any one of many different cell types

Polymer a large molecule made up of a number of smaller molecules linked together

Polymerase An enzyme that catalyses the synthesis of nucleic acids

Polypeptide a large chain of amino acids linked by peptide bonds

Primary germ layers embryonic cells in which the process of differentiation is just beginning

Protein a molecule made up of polypeptide chains

Protein synthesis the mechanism that produces the various proteins in our cells and tissues.

Proteomics the study of the structure and function of proteins

Proto-oncogene A gene that may be converted into an *oncogene* by mutation

Receptor a protein on the cell membrane that binds to a particular hormone or other chemical, initiating the hormone's action on the cell

Recessive gene a gene that exerts its full effect only when two copies are present

Regulatory gene a gene involved in the regulation of expression of another gene

Regulatory protein a protein that has a role in differential gene expression or metabolic control

Ribosome an organelle that is the site of protein synthesis

RNA ribonucleic acid, a nucleic acid of a specific sequence

Single gene disorder a disease caused by a mutation in a single gene

Somatic cells all body cells, except the germ line cells

Stem cell an undifferentiated cell; that is, one that has the potential to become one of the body's many different cell types

Structural genomics the analysis of the molecular structure of proteins

Taxonomy the classification of living things

Terminator genes in genetically modified crops, genes that prevent the plant from setting seed

Therapeutic cloning the process of creating an embryo by nuclear transfer, with the object of obtaining cells that can be transplanted into the donor to treat a disease without being rejected by the donor's immune system

Thyroid gland a gland in the throat important in regulating metabolism

Transcription the enzymatic process that allows the genetic information contained in DNA to specify a complementary sequence of bases in messenger RNA

Transfer RNA a class of small RNA molecules that act at the first step in protein synthesis, bringing individual amino acids to their correct positions

Transgenic organism an organism whose DNA has been modified by the introduction of genetic material from another species

Translation the process in which the genetic information in an RNA molecule specifies the sequence of amino acids during protein synthesis

Virus a self-replicating, infectious nucleic acid–protein complex that can replicate in host eucaryote cells

X-linked disorder a disease caused by a faulty gene located on the X chromosome

FURTHER READING

Introduction

J Watson and A Berry (2003) *DNA: the secret of life*, AA Knopf, New York

R Dawkins (1995) *River out of Eden*, Weidenfeld & Nicholson, London

Chapter 1

SF Gilbert (2003) *Developmental biology*, Sinauer, New York

C Masters & R Holmes (1975) *Haemoglobin, isoenzymes and tissue differentiation*, Elsevier, New York

W Murrell, C Masters, R Willis & D Crane (1994) 'On the ontogeny of cardiac gene transcripts', *Mechanisms of Ageing and Development*, 77, 109–126.

DNA and protein synthesis
web.jjay.cuny.edu/~acarpi/nsc/12-dna.htm

Forensic fact file: DNA profiling
www.nifs.com.au/factfiles/DNA

Chapter 2

The DNA diagnostic business
www.biz-lib.com/zbub141.html

Your genes, your health
Yourgenesyourhealth.org/-10k-1 Nov 2004

These chips are good for health
www.cbsnews.com/stories/2002/04/14/health/main506124.shtml

Chromosome FAQ
www.ornl.gov/sci/techresource/Human_Genome/posters/chromosome/faqs.shtml

Chapter 3

K Davies (2000) *Cracking the genome*, Simon & Schuster, New York
M Ridley (1999) *Genome*, Perennial, London
MA Rothstein (ed) (1999), *Genetic secrets: protecting privacy and confidentiality in the genetic era*, Yale University Press, New Haven
Blueprint of the body
www.cnn.com/specials/2000/genome

Chapter 4

G Kolata (1997) *Clone*, Penguin, London
I Wilmut, K Campbell & C Trudge (2000) *The second creation*, Farrar, Strauss & Giraux, London
EM De Robertis & JB Gurdon (1979) 'Gene transplantation and the analysis of development', *Scientific American*, 241(6), 74–82
LR Kass & JQ Wilson (eds) (1998) *The ethics of human cloning*, AEI Press, Washington

Chapter 5

MW Fox (1999) *Beyond evolution: the genetically altered future of plants, animals, the earth and humans*, Lyons Press, New York
What is genetic engineering? a simple introduction
www.geneticengineering.org/dna6/default.htm
Genetic engineering
en.wikipedia.org/wiki/Genetic_engineering

Chapter 6

F Guilak, DL Butler, SA Goldstein & D Mooney (eds) (2003) *Functional tissue engineering*, Springer-Verlag, New York
Tissue engineering: a biological solution for tissue damage, loss or end stage organ failure
ibj.pasteur.ac.ir/ibj7/Haydaran.htm
Human embryonic stem cells may promise medical advances
www.eurekalert.org/pub_releases/2004-02/aaft-hes020504.php

Chapter 7

B Alberts et al (2003) *Molecular biology of the cell*, Garland, New York

C De Duve (1984) *A guided tour of the living cell*, Scientific American, New York

C Masters (1996) 'On the role of the cytoskeleton in metabolic compartmentation' in JE Hesketh & IF Pryme (eds) *The Cytoskeleton,* vol 2, 1-30, JAI Press, London

C Masters & D Crane (1995) *The peroxisome: a vital organelle*, Cambridge University Press, Cambridge

The DNA, RNA and proteins
Bioinformatics.org/tutorial/1-1.html

Chapter 8

W Lu et al (2004) 'Gene regulation and DNA damage in the ageing human brain', *Nature* 4299(6994) 883–891.

C Masters (2000) 'Ageing. degenerative disorders and peroxisomes', Proceedings of the 6th Asia/Oceania Congress of Gerontology, pp. 31–33

Ageing process
www.healthandage.com/html/res/primer2.htm

Ageing research
www.abdn.ac.uk/~nhi158/ageing.htm

How do DNA damage and repair relate to cancer
www.infoaging.org/b-dna-5-cancer.htm

Chapter 9

R Gosden (2000) *Designing babies: the brave new world of reproductive technology*, Freeman, San Francico

C Masters & D Crane (1998) 'On the role of the peroxisome in cell differentiation and carcinogenesis', *Molecular and Cellular Biochemistry*, 187:85–97

Fetal alcohol syndrome
www.well.com/user/woa/fsfas.htm

Breast cancer susceptibility genes play a role in DNA repair
www.eurekalert.org/pub_releases/2003-11/twi-bcs111703.php

In vitro fertilisation
www.fact-index.com/i/in/in_vitro_fertilisation_1.html

Chapter 10
S Hanash (2003) 'Disease proteomics', *Nature* 422, pp. 226–232.
Proteonomics
www2.carthage.edu/~pfaffle/hgp/proteonomics.html
The secret of life
www.time.com/time/covers/1101030217/
Proteonomics
www.biology.eku.edu/FARRAR/gen-prot.htm
Cancer biomarkers
www.clinchem.org/cgi/content/full/48/8/1147

Chapter 11
J Rifkin (1998) *The biotech century*, Phoenix, London
RC Lewontin (1993) *The doctrine of DNA*, Penguin:London
DG Springham & V Moses (eds) (2002) *Biotechnology: the science and business*, Gordon & Breach, New York
DNA insights will transform science and medicine
www.mercola.com/2001/24/human_genome.htm
Junk DNA yields a new kind of gene
www.eurekalert.org/pub_releases?2004-06/hms
Will future computers be made of DNA?
Library.thinkquest.org/3037/dna.htm
The future of DNA
www.gene.ch/gentech/1997/8,96-5,97/msg00225.html
Dr DNA M.D.
www.chemistry.org/portal/a/c/s/i/feature
C Masters (1985) 'Perspectives on prions', *Nature* 314, pp. 15–16.

INDEX